Defusing the Toxics Threat: Controlling Pesticides and Industrial Waste

Sandra Postel

Worldwatch Paper 79
September 1987

Library of Congress Catalog Card Number 87-051021
ISBN 0-916468-80-1

Printed on recycled paper

Table of Contents

Introduction 5
Shadows of the Chemical Age 7
Consequences and Risks 15
Breaking the Pesticide Habit 25
Rethinking Industrial Waste Management 36
Detoxifying the Environment 46
Notes 57

In 1948, Swiss chemist Paul Müller received the Nobel Prize in medicine for discovering the insecticidal properties of DDT. By then, the insect killer had saved untold lives from malaria, typhus, and other deadly diseases. To many people working during World War II to protect Allied troops from these insect-borne epidemics, DDT was a miracle chemical. One British entomologist wrote in 1942 that "the new insecticide appeared to be so exactly what we wanted that it looked too good to be true."[1]

But after little more than a generation, this much-heralded substance had been banned or severely restricted from use in many countries. It had spread throughout the globe, contaminated the food chain, pushed bald eagles and other predatory birds toward extinction, and accumulated in fish, wildlife, and people. Ugly surprises had cast dark shadows over DDT's many blessings.

This uneasy counterpoint of benefits and risks marks the course of the chemical age. Events continue to reveal that "better living through chemistry" comes with serious costs, some of which have only recently come to light. Pesticides thought to degrade in soils turn up in rural drinking water wells. Underground plumes of toxic chemicals emanating from abandoned waste sites contaminate city water supplies. A gas leak at a chemical production plant in Bhopal, India, kills more than 2,000 people. Pesticides spilled into the Rhine River from a warehouse near Basel, Switzerland, destroy a half million fish, disrupt water supplies, and cause considerable ecological damage. In many ways—some dramatic, others insidious—chemicals seem to be escaping society's control.

Many thanks to Angela Coyle, Gretchen Daily, Lori Heise, and Susan Norris for the help they so cheerfully provided at various stages of this paper. I extend thanks also to Kim Christiansen, Gary Davis, Michael Hansen, Maureen Hinkle, Banpot Napompeth, Katherine Reichelderfer, Roger Schecter, and Graham Searle for their helpful comments on a draft of the manuscript.

The use of pesticides in agriculture and the discarding of industrial chemical waste into the air, soil, and water constitute two major pathways of human exposure to toxics. These practices release hundreds of millions of tons of potentially hazardous substances into the environment each year. In the absence of complete data on the health and ecological effects of the vast majority of chemicals in use, we can only speculate about the long-term consequences. But isolated cases of contamination and recent epidemiological studies give clear cause for concern.[2]

Strategies that reduce pesticide use in agriculture and that minimize waste generation in industry offer cost-effective approaches to lessening risks from toxics. They differ markedly from current practices, requiring new ways of thinking. The quick fixes of pesticide spraying and end-of-pipe pollution control are replaced with new production systems aimed at reconciling economic profits and environmental protection. In agriculture, for example, rather than automatically applying pesticides to maximize yield from a monoculture cropping system, farmers consider whether crop rotation, intercropping, or a biological control agent might allow them to sustain profits while protecting groundwater from contamination. Similarly, industrial managers explore whether a new manufacturing process, different raw materials, or a new product altogether could save money by reducing hazardous waste rather than treating and disposing of whatever is generated.

With technologies and methods now available, pesticide use could probably be halved and the creation of industrial waste cut by a third or more over the next decade. Successful efforts to date suggest that farmers and manufacturers would benefit economically, while people and the environment would receive better protection. But existing policies fail to promote these new techniques; in some cases, they actually undermine them. Many Third World governments, for instance, heavily subsidize pesticide use while investing little in safer, more ecologically based pest control methods. Virtually all governments indirectly subsidize the disposal of industrial waste by failing to make companies pay the full social costs of their practices. Moreover, many farmers and manufacturers simply do not know about alternative methods, or cannot afford to risk trying them.

"Strategies that reduce pesticide use and that minimize waste generation aim at reconciling economic profits and environmental protection."

In recent years, a few countries have taken steps to reduce reliance on pesticides or to promote industrial waste reduction and recycling. Overall, however, public commitments to research and development, demonstration projects, training, and education in these methods are woefully inadequate to bring them into widespread use. These strategies provide a real chance of getting toxics under control, and the time is ripe to seize that opportunity.

Shadows of the Chemical Age

Modern industry uses copious quantities of chemicals in the manufacture of steel, paper, wood products, and other everyday materials. With the quadrupling of global economic activity since 1950, materials production rose markedly, as did the amount of chemical waste created. Over the same period, laboratory scientists transformed the manufacture of synthetic organic chemicals from a tiny, specialty enterprise to a ubiquitous, powerful industry.

Organic compounds, by definition, are those containing carbon. Unique among chemical elements, carbon easily bonds with itself to form chains and rings, and can combine in various ways with other common elements, including hydrogen, nitrogen, and chlorine. During recent decades, laboratory scientists learned to exploit carbon's properties not only to recreate chemicals found in nature, but also to fashion millions of entirely new compounds that have no natural analogs. In so doing, they paved the way for a host of new industrial and consumer products that have greatly changed the shape of society—from plastics and pesticides to birth control pills and polyester fibers.[3]

Both the volume and number of manufactured chemicals have burgeoned since World War II. In the United States, annual production of synthetic organic chemicals rose 15-fold between 1945 and 1985, from 6.7 million metric tons to 102 million. (See Figure 1.) This growth far outpaced that of the nation's overall economic activity. In 1945, some 4.5 kilograms of synthetic organic compounds were produced for every $1,000 of gross national product; by 1979, this figure had climbed to 30 kilograms (in real terms), more than a sixfold increase.

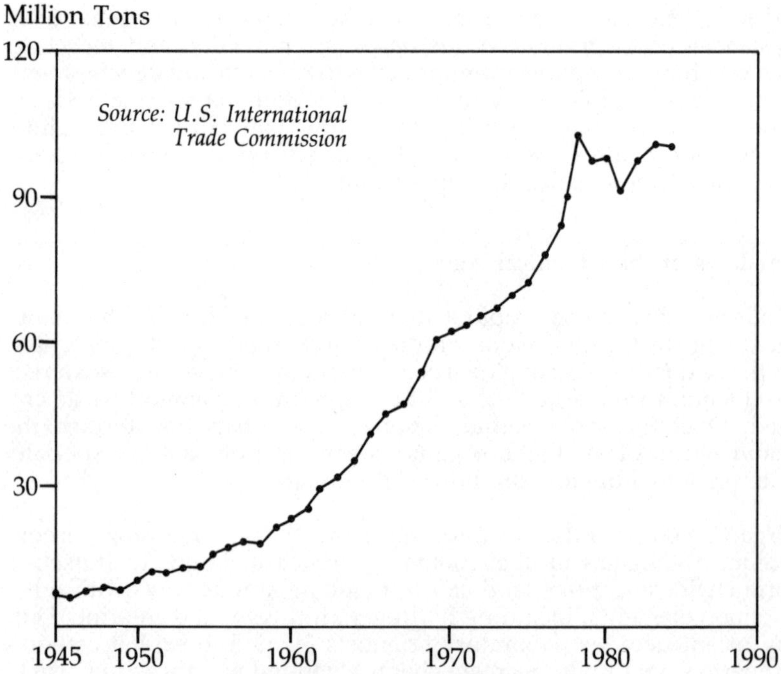

Figure 1: Synthetic Organic Chemicals Production, United States, 1945-85

Worldwide, some 70,000 chemicals are presently in everyday use, with between 500 and 1,000 new ones added to the list each year. No limit to the number of possible syntheses is in sight.[4]

Prior to the forties, farmers relied on a combination of mechanical, chemical, and biological methods to limit pest damage to crops. The discovery of DDT, however, ushered in an era of almost exclusive dependence on chemicals for pest control. DDT was safer and more effective than the arsenic, heavy metal, cyanide, and nicotine compounds that had long been used. It was relatively inexpensive, re-

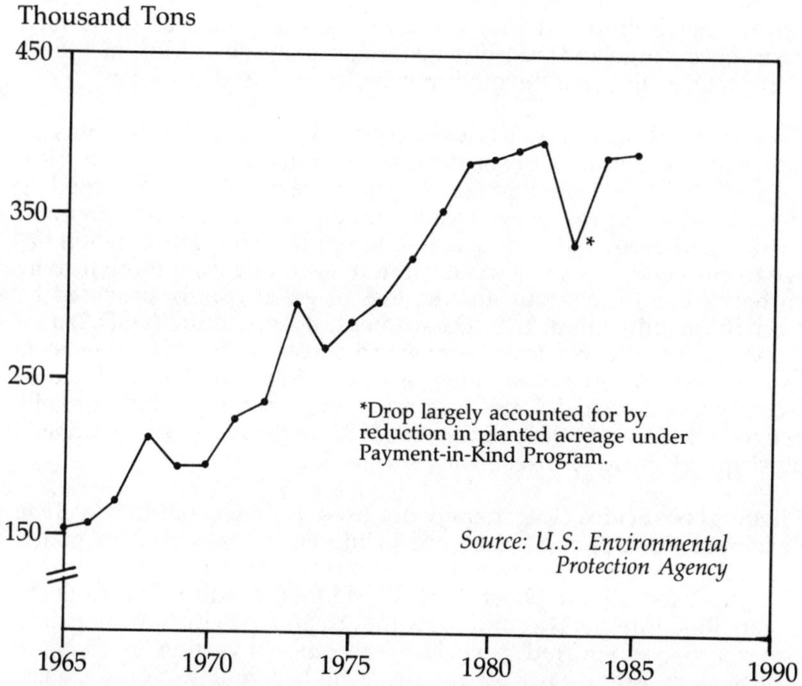

Figure 2: Pesticide Use in Agriculture, United States, 1965-85

mained active a long time in the soil, and was toxic to a broad spectrum of insects. Synthetic chemicals freed farmers from much of the worry and complexity of controlling insect pests. Demand for pesticides skyrocketed, and interest in nonchemical methods of pest control dwindled.[5]

In the United States, pesticide use in agriculture nearly tripled between 1965 and 1985. (See Figure 2.) Farmers applied 390,000 tons of pesticides to the nation's agricultural land in 1985, an average of about 2.8 kilograms (6.2 pounds) per hectare planted. Roughly 70 percent of all cropland (not counting land in alfalfa or other hay, pasture, or rangeland) receives some dosage of pesticides, including 95 percent of the area devoted to corn, cotton, and soybeans.[6]

Greater use of herbicides for weed control has accounted for much of the recent growth in pesticide use in many industrial countries. Farmers striving to maximize output increasingly have turned to chemicals to suppress weeds that compete with crops for water and nutrients, thereby reducing yields. Those practicing conservation tillage to curb energy use or soil erosion tend to rely even more heavily on herbicides to make up for the loss of weed control provided by mechanical cultivation. U.S. Department of Agriculture (USDA) analysts expect herbicides to account for 85 percent of total U.S. pesticide use in 1987. Whereas the value of insecticide sales has declined in recent years, herbicide sales rose an average of 5.4 percent annually between 1972 and 1984. Similarly, 1982 herbicide sales in Canada outstripped those of insecticides sevenfold.[7]

Chemical pesticides are generally not used as widely or intensively in developing countries as they are in industrial ones. But in many, usage has been growing rapidly. Chemicals were part of the package of inputs promoted to boost Third World food production during the Green Revolution. The shift toward greater production of export crops also has spurred pesticide use, since investing in chemical inputs more often pays off for these higher valued crops. Consequently, pesticide use has risen substantially in much of the Third World over the last few decades. In India, for example, use increased from about 2,000 tons annually in the fifties to more than 80,000 tons in the mid-eighties. Some 80 million hectares of India's cropland now receive treatment with chemical pesticides, compared with just 6 million in 1960.[8]

Unfortunately, reliable data on pesticide use are lacking for most countries and regions. The value of pesticide imports provides the best available surrogate for discerning global and regional trends. Worldwide, pesticide imports rose 2.6-fold (in real terms) between 1972 and 1984, to $5.3 billion. (See Table 1.) Imports quadrupled in the Soviet Union and more than tripled in Asia and North America during this period. Many developing countries, besides stepping up

Table 1: Value of Pesticide Imports, World and by Region, 1972-84

Region	1972[1]	1984[2]	Change
	(million 1985 dollars)		(percent)
Soviet Union	132	552	+318
North America	142	535	+277
Asia	314	1,132	+261
Europe	824	2,014	+144
Africa	269	522	+ 94
Oceania	30	47	+ 57
Latin America	340	503	+ 48
World	2,051	5,305	+159

[1]Average over 1971-73.
[2]Average over 1983-85.

Source: U.N. Food and Agriculture Organization, *Trade Yearbooks* (Rome: 1977 and 1985).

imports, are expanding their capacities to produce pesticides domestically, so total usage undoubtedly is on the rise.[9]

Concerning chemical waste, data on generation and disposal are much sketchier and more confusing than those for pesticides. Countries apply different definitions to what they variously call "hazardous," "special," or simply "industrial" waste, which obscures comparisons across countries. Official figures on waste often reflect legal definitions more than they do the actual volume of dangerous by-products discharged to the environment. In the United States, for example, toxics released to waterways are regulated under the nation's clean water law, not under its hazardous waste law, and so are not included in official estimates of hazardous waste. Finally, no country has a long-term historical record of hazardous waste generation and disposal because most only began distinguishing hazardous from other waste within the last 10–15 years.

Estimates of the total amount of hazardous waste generated in the United States have varied greatly. Applying a somewhat broader definition than that in the law, the Congressional Budget Office (CBO) placed the figure at about 266 million tons for 1983, more than one ton for every American. With the aid of a model incorporating basic industrial output and process data, the CBO also derived a picture of which industries were generating the most waste: Those in the category of chemicals and allied products were found to produce nearly half the total 266 million tons, with those in primary metals accounting for 18 percent, and those in petroleum and coal products, 12 percent.[10]

Waste streams from the organic chemical industry alone amounted to 47 million tons, 18 percent of the total, making it the largest single generator. In addition, the Environmental Protection Agency (EPA) ranks 80 percent of the organic chemical industry's waste streams as 9 or 10 on an increasing, relative toxicity scale of 1 to 10. So not only does this industry produce the greatest quantity of hazardous waste, but its waste is among the most highly toxic of all.[11]

Current methods of managing industrial waste in many cases still reflect the "out of sight, out of mind" mentality of the fifties and sixties. The CBO estimates that two-thirds of the hazardous waste produced in the United States is disposed of in or on the land through the use of injection wells, pits, ponds and lagoons, or landfills. (See Table 2.) Each of these practices risks contaminating groundwater, since experts claim that even the most carefully constructed landfill or surface impoundment will eventually leak. Although legislation designed to curb land disposal of especially hazardous waste was passed in 1984, EPA has delayed its implementation because of a shortage of facilities to treat and handle the waste in safer ways. About another one-fifth of U.S. hazardous waste gets discharged to sewers or directly into rivers and streams. Only a small fraction is recycled, destroyed, detoxified, or otherwise rendered harmless before being released into the environment.[12]

With their higher population densities, most West European countries have tended to rely somewhat less on land disposal methods. Compared with the United States, larger shares of waste are either treated prior to disposal or incinerated on land or at sea. Of the waste handled in West Germany, for example, an estimated 15 percent gets

"Experts claim that even the most carefully constructed landfill or surface impoundment will eventually leak."

Table 2: Hazardous Waste Management Methods, United States, 1983

Management Method	Share of Total Waste Managed
	(percent)
Land disposal[1]	67
Discharge to sewers, rivers, streams	22
Distillation for recovery of solvents	4
Burning in industrial boilers	4
Chemical treatment by oxidation	1
Land treatment of biodegradable waste	1
Incineration	1
Recovery of metals through ion exchange	—[2]
Total	100

[1]Includes injection wells (25 percent of total), surface impoundments (19 percent), hazardous waste landfills (13 percent), and sanitary landfills (10 percent).
[2]Less than 1 percent.

Source: U.S. Congressional Budget Office, *Hazardous Waste Management: Recent Changes and Policy Alternatives* (Washington, D.C.: U.S. Government Printing Office, 1985).

incinerated, 35 percent receives treatment for detoxification, and 50 percent is disposed of in landfills. An important exception, however, is the United Kingdom, where hazardous waste and domestic waste are discarded together in landfills that have been likened to complex biochemical reactors. This practice, called "co-disposal," is generally viewed with great skepticism by others in Europe. British authorities consider it safe, as long as landfill operators control the wastes entering their site and carefully monitor the landfill's performance. But a 1986 report by the U.K.'s Hazardous Waste Inspectorate painted a grim picture of actual practices, and concluded: "If we have avoided major problems with co-disposal landfill in the UK, the Inspectorate considers that in some cases this is due more to luck than judgement."[13]

Much hazardous waste crosses national borders because of insufficient domestic disposal capacities and the lure of cheaper disposal sites elsewhere. Some 200,000–300,000 tons annually are shipped from Western Europe into Eastern Europe, where at least 90 percent of all hazardous waste is disposed of on land. The Danes export about 10 percent of their hazardous waste, and the West Germans more than a quarter of theirs, with much of it bound for a large landfill just across the border in East Germany.[14]

Many developing countries now industrializing their economies are generating growing volumes of hazardous waste, though the totals—impossible to specify—are still dwarfed by those in the West. Few have implemented regulations controlling this waste, and even fewer have the advanced technologies needed to do so adequately. H. Jeffrey Leonard of The Conservation Foundation in Washington, D.C., writes that "such exploding urban areas as Mexico City, São Paulo, Seoul, Jakarta, Lagos, Lima, and Calcutta show that the most ominous examples of serious environmental contamination in the world are found neither in the heavily industrialized countries nor in the poorest of the poor countries, but rather in and around the cities of those countries that have recently experienced rapid industrial development and urban growth."[15]

The 400 kilometers between Rio de Janeiro and São Paulo, along Brazil's Atlantic Coast, now constitute one of the most heavily industrialized regions in the world. Chronically high levels of industrial pollution have given the town of Cubatão the menacing nickname "Valley of Death." The unregulated dumping of waste by several hundred metallurgical and other factories into Sepetiba Bay—Rio de Janeiro's primary source of seafood—is believed responsible for high levels of chromium, zinc, and cadmium found in its shellfish.[16]

Similarly, many industries in Mexico City discharge wastewater contaminated with heavy metals and toxic organic compounds into the city sewer system. After little or no treatment, much of this wastewater is pumped to agricultural areas for use in irrigation. Officials have discovered contaminants in vegetables and other crops, raising concerns about long-term health risks to consumers.[17] For this chronically water-scarce city, reusing wastewater appears essential, but failure to control toxic pollution threatens the viability of this strategy.

India now has an estimated 4,000 chemical factories. Between 1970 and 1980, production of pesticides increased 13-fold, dye and pigment production more than doubled, and the output of organic chemicals overall rose 41 percent. These and other industries generate large quantities of toxic waste; most of it goes to landfills or directly to rivers and streams with little or no treatment. China produces some 400 million tons of industrial waste and tailings annually, much of it undoubtedly hazardous. Most is dumped on the outskirts of cities or released into surface waters. Mounds of potentially harmful waste reportedly occupy some 60,000 hectares of China's land today.[18]

The litany of neglect seems almost endless. In densely populated Hong Kong, most hazardous waste is discharged to sewers or sent to landfills designed only for nontoxic municipal waste. Malaysia has no facilities capable of handling hazardous waste. Yet a 1985 survey of 700 industries along Peninsular Malaysia's heavily industrialized west coast revealed that 85 percent of them were generating hazardous waste. Industries either send it to unsecured domestic landfills, stockpile it, or dump it indiscriminately into the environment.[19]

Consequences and Risks

The hazard to health posed by a chemical after it enters the environment depends primarily on two factors: its toxicity and the extent of human exposure to it. Unfortunately, knowledge of the harmful effects of synthetic organic compounds has lagged far behind their introduction to the marketplace. The U.S. National Research Council (NRC) estimates that no information on toxic effects is available for 79 percent of the more than 48,500 chemicals listed in EPA's inventory of toxic substances. Fewer than a fifth have been tested for acute effects, and fewer than a tenth for chronic (for example, cancer-causing), reproductive, or mutagenic effects. Pesticides generally have received more extensive testing, but there, too, serious gaps remain. By allowing the production and release of these compounds without understanding their damaging effects, society has set itself up for unpleasant surprises.[20]

16 Pesticides account for only a small share of the 70,000 chemicals in common use, but they pose some of the greatest potential hazards. Unlike most industrial compounds, pesticides are purposely designed to alter or kill living organisms. Moreover, since they are spread widely over the land, they pose risks not only to farm workers but to the general population through residues in food crops and through contamination of drinking water.

Between 400,000 and 2 million pesticide poisonings occur worldwide each year, most of them among farmers in developing countries. The 10,000–40,000 such poisonings that are thought to result in death each year dwarf the 2,000 deaths caused by the toxic gas leak at the pesticide manufacturing plant in Bhopal, India, tragic though that accident was. No comparable estimates exist for deaths and disease caused by chronic, low-level exposures to farm chemicals, but the picture is far from comforting.[21]

Many older chemicals that industrial countries have restricted or outlawed are still widely used by farmers in the Third World. DDT and benzene hexachloride (BHC), both banned from use in the United States and much of Europe, account for about three-quarters of total pesticide use in India. One analysis there of cereals, eggs, vegetables, and other foods found that 30 percent had pesticide residues exceeding tolerance limits set by the World Health Organization (WHO). Residues of DDT and BHC, both suspected carcinogens, were found in all 75 samples of breast milk collected from women in India's Punjab region. Through their mothers' milk, babies daily were ingesting 21 times the amount of these chemicals considered acceptable. Similarly, samples of breast milk from Nicaraguan women have shown DDT levels an astounding 45 times greater than WHO's tolerance limits.[22]

Ironically, through imported foods consumers in industrial countries remain exposed to these chemicals even though their own governments may have restricted or banned them from domestic use, completing what some have called a "circle of poison." In some countries,

"Samples of breast milk from Nicaraguan women have shown DDT levels an astounding 45 times greater than WHO's tolerance limits."

including the United Kingdom, for example, regulatory agencies do not routinely or systematically monitor pesticide residues in food. A monitoring program exists in the United States, but the General Accounting Office estimates that less than 1 percent of imported fruits and vegetables is inspected for pesticides banned domestically. In 1983, the Natural Resources Defense Council independently tested Latin American coffee beans sold in New York City markets, and in each of four samples found multiple residues of DDT, BHC, and other persistent pesticides. The Brazilian beans, the most popular variety in the United States, contained residues of five suspected carcinogens.[23]

A May 1987 study by the National Research Council suggested that pesticide residues in domestically grown foods could also add to the nation's cancer risk. In its worst-case estimate, the NRC study calculated an increased risk of 5,800 cancer cases per million people over a 70-year lifetime, far higher than the 1 per million "acceptable" risk level that EPA often applies to cancer hazards. The NRC figure translates to roughly 1.4 million additional cases for the current U.S. population—or 20,000 additional cases per year. Nearly 80 percent of the estimated risk derived from just 15 foods, with tomatoes, beef, potatoes, oranges, and lettuce leading the list.[24]

A third pathway of pesticide exposure—contaminated drinking water—is of rapidly growing concern. No nation has systematically monitored its water supplies for pesticides, so the full extent of contamination is unknown. Yet, again, the evidence available points toward some serious problems.

In the United Kingdom, preliminary survey results suggest widespread contamination of rivers and streams in the agricultural areas of eastern England. The herbicide atrazine, a suspected carcinogen, contaminates most surface waters in the region and has been found at levels nearly three times the maximum acceptable concentration for herbicides in drinking water set by the European Community.[25]

In the United States, routine agricultural practices have contaminated groundwater with 17 different pesticides in at least 23 states. The nation's two most widely used herbicides—alachlor and atrazine—

were among the pesticides most frequently detected. Tests have shown alachlor to cause cancer in laboratory animals, making it a probable human carcinogen. The states with the best monitoring programs (California, New York, and Iowa) have found the greatest number of pesticides, suggesting that more extensive monitoring in other states would reveal a more pervasive problem. Iowa's surveillance efforts indicate that more than a quarter of Iowans use drinking water contaminated with pesticides.[26]

EPA plans to analyze groundwater from some 1,500 wells between 1987 and 1990 to better uncover the magnitude of the problem nationwide. Meanwhile, USDA researchers Elizabeth Nielsen and Linda Lee have attempted a broad-brush assessment of the potential risks of pesticide contamination of groundwater. They examined county-level hydrogeologic data along with figures on pesticide use and determined that about a third of all counties in the contiguous 48 states are vulnerable to contamination. Major regions of high contamination potential include the Atlantic Coastal Plain, the Mississippi Delta, the northern Corn Belt, and California's Central Valley. Their findings match to a great extent the locations of known incidents of groundwater contamination by pesticides.[27]

Based on the study results, nearly 50 million U.S. residents—most of them in rural areas—are potentially at risk of exposure to pesticide-contaminated groundwater. More than 17 million people get their drinking water from private wells in these high-risk regions. Very little of this private water is treated or monitored. An additional 29 million people in these areas use groundwater supplied through community systems. Though the federal government regulates public water suppliers, it has set drinking water standards for only 6 pesticides—none of which are among the 17 that EPA reports have contaminated groundwater.[28]

The relative threat posed by pesticide poisonings, food residues, and contaminated drinking water varies with the type of pesticide used and the care taken during application. Organochlorine or DDT-type insecticides are not very acutely toxic. But their persistence, along with their ability to accumulate in fatty tissue, has led to their buildup in the food chain and to the high concentrations found in breast milk.

"Insects and weeds now reduce crop production by about 30 percent, apparently no less than before the chemical age dawned."

Organophosphates degrade more readily, but are more acutely toxic. Many Third World farmers lack the training and equipment needed to apply such highly toxic chemicals safely, or cannot read well enough to understand the label instructions. A 1985 survey in one county of the Brazilian state of Rio de Janeiro found that 6 out of 10 farmers using pesticides had suffered acute poisonings, two-thirds of them from organophosphates. Finally, many of the modern herbicides exhibit strong potential for leaching to groundwater, as the U.S. experience shows. Thus, the risks, while almost always present, vary in form and degree.[29]

Presumably greater pesticide use is justified if the benefits outweigh the costs and risks. But this case is getting harder to make. Insects and weeds now reduce crop production by about 30 percent, apparently no less than before the chemical age dawned. Because of stricter regulatory requirements and the greater complexity of modern chemicals, industry now spends $20–45 million bringing a new pesticide to market, compared with about $1.2 million in 1956.[30]

More importantly, chemicals no longer provide the effective means of crop protection they once did. In response to heavier pesticide use, pests have evolved mechanisms of detoxifying and resisting the action of chemicals designed to kill them. In 1938, scientists knew of just seven insect and mite species that had acquired resistance to pesticides. By 1984, that figure had climbed to 447, and included most of the world's major pests. (See Figure 3.) Resistance in weeds was virtually nonexistent before 1970. But since then, with the growth of herbicide use, at least 48 weed species have gained resistance to chemicals.[31]

Farmers and pesticide producers have thus locked themselves into a race with the rapid evolution of crop pests. Chemicals intended to enhance and stabilize agricultural production have in some cases done just the opposite. In northeastern Mexico, a major pest of cotton—the tobacco budworm—developed resistance to every registered insecticide. Mounting crop damage caused the area planted in cotton during the sixties to drop from more than 280,000 hectares to a mere 400 hectares. Little cotton is grown there today. Similarly, in Nicaragua, 15 years of heavy insecticide use on cotton were followed

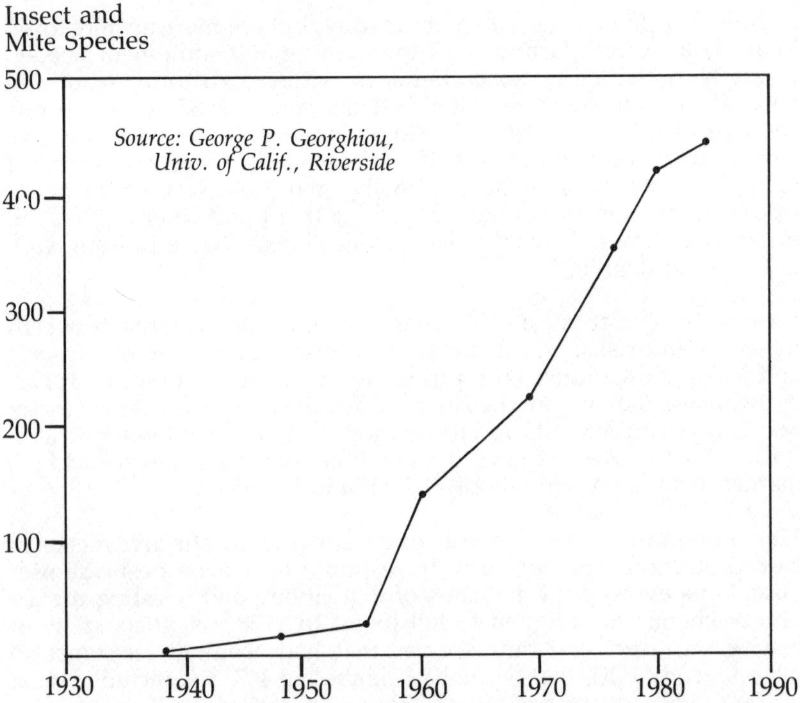

Figure 3: Number of Insect and Mite Species Resistant to Pesticides, 1938-84

by four years in which yields fell by 30 percent. Pests had acquired resistance, the chemicals had killed pests' natural enemies, and new pests had emerged. As crop damage increased, desperate farmers reacted by applying more insecticide, which only aggravated the problem. In a classic case of the "pesticide treadmill," insect control costs rose to a third of total cotton production costs.[32]

In Suffolk County, Long Island—the leading farm county in New York State—chemicals are losing the battle against the Colorado po-

"Most countries still have only a vague idea of the magnitude of air, water, and soil contamination caused by industrial chemicals."

tato beetle. The beetle has acquired resistance to all major insecticides registered for use on potatoes. Growers spray up to 10 times per season, and pest control costs have climbed as high as $700 per hectare. Meanwhile, heavy application of pesticides has caused extensive contamination of groundwater, the region's sole source of drinking water.[33]

Other cropping systems at risk from resistance problems include cabbage and rice in Southeast Asia, corn in the United States, potatoes in parts of Europe as well as the eastern United States, sugar beets in the United Kingdom, and cotton in many other parts of the world. According to entomologist George P. Georghiou, resistance unquestionably "poses a serious obstacle to efforts to increase agricultural production." While an entrenched agrochemicals industry continues to propound the virtues and necessity of reliance on pesticides, the facts cry out for new solutions to pest problems.[34]

As with pesticides, the consequences and risks from the burgeoning use of industrial chemicals have only begun to be characterized. Nearly a decade has passed since the Love Canal site in Niagara Falls, New York, spotlighted the insidious hazards posed by indiscriminate chemical waste disposal. Yet most countries still have only a vague idea of the magnitude of air, water, and soil contamination caused by industrial chemicals.

Tens of thousands of active and abandoned waste disposal sites dot the landscapes of industrial countries. Corrosive acids, persistent organics, and toxic metals accumulated for decades with little thought about whether they would enter the environment. Preliminary estimates by the West German länder (states) suggest that as many as 35,000 problem sites exist nationwide. Efforts to assess how many pose serious risks are still going on, but officials expect corrective actions to require at least 18 billion German marks ($9.7 billion) over the next decade. In Denmark, which, like West Germany, relies heavily on groundwater, up to 2,000 sites are believed to be contaminated. Officials currently anticipate cleanup expenditures of some 1 billion Danish kroner ($143 million).[35]

Perhaps because of the jolting impact of Love Canal, efforts to assess contamination from waste disposal practices in the United States are further along than in most European countries. An EPA study completed in the early eighties found that more than 70 percent of the 80,000 pits, ponds, and lagoons containing hazardous chemicals did not have liners to guard against seepage. The geologic settings of nearly half the sites were such that any seepage that did occur would reach groundwater fairly quickly. All factors considered, 72,000 impoundments—90 percent of the total—are thought to pose some threat of groundwater contamination.[36]

As of July 1987, EPA had placed 951 landfills, impoundments, and other waste sites on its National Priority List, which covers sites needing urgent attention. The agency estimates that the list will grow to not more than 2,500 sites and that cleanup costs may total some $23 billion. The Congressional Office of Technology Assessment (OTA) estimates, however, that the number of priority sites could climb to 10,000 and expenditures to $100 billion, roughly $400 for every U.S. resident. Legislation enacted in October 1986 established an $8.5-billion, five-year program (the Superfund) to step up progress on site cleanups. While a good start, this commitment obviously falls short of total needs.[37]

Meanwhile, no clear picture has emerged of the extent of groundwater contamination nationwide. Drilling monitoring wells and analyzing water samples are time-consuming and expensive. Yet more than 200 substances have been identified in the nation's groundwater, including 175 organic chemicals. Thirty-two of these organics (including some pesticides) and five metals are known or suspected carcinogens. Equally unsettling, a substantial share of the contaminants frequently found have not even been tested for long-term health effects. Most remain unregulated and unmonitored: EPA has set drinking water standards for only about two dozen of the hundreds of substances detected in groundwater.[38]

Chemical wastes are not purposely spread over cropland the way pesticides are, but they can contaminate food nonetheless. Some drift through the air to lakes and farmlands, contaminating fish, crops, and grazing livestock. Toxic discharges to rivers, bays, and estuaries contaminate commercial fisheries. For some populations, contaminated food adds appreciably to the risk of toxic exposure. Heavy metals and chlorinated organics pose the greatest concerns because many of these substances can accumulate in the food chain.

According to a joint Canadian-U.S. study, people living in the Great Lakes region have greater exposure to toxic chemicals than most other North Americans, largely because of contaminants in the food they consume. Fish caught in the Great Lakes have toxic residues frequently exceeding allowable levels. Studies of polychlorinated biphenyls in breast milk from women in the region indicate that the daily exposure of some infants to this toxic organic exceeds federal standards. Similarly, in Sweden authorities have found high levels of dioxin, a very toxic chlorinated organic, in fish from the Baltic Sea and in crabs along the west coast. According to one environment official, WHO is considering setting a daily intake standard for dioxin that would take most of these crabs off the market.[39]

Researchers in Poland have found alarmingly high concentrations of heavy metals in vegetables in the heavily industrialized region of Upper Silesia. This area harbors numerous smelters and metals factories that release toxic metals to the air and to unsecured dumps. Soil samples taken from vegetable gardens in Upper Silesia have contained levels of cadmium, mercury, lead, and zinc between 30 and 70 percent higher than levels considered safe by WHO. Inhabitants of this region reportedly experience 30 percent more cancer cases than the average for Poles; children more frequently show signs of mental retardation.[40]

At the same time that countries grapple with the consequences of past practices, current methods of managing chemicals are adding to the total risks. A large share of a growing volume of waste still gets disposed of on or in the land in most countries, and, as mentioned, all landfills are expected to leak eventually. Incinerators can destroy a good portion of organic wastes, but concern and controversy remains over their potential release of dioxins. Incineration that meets strict temperature, residence time, and other operating standards apparently can alleviate the dioxin threat and assure virtually complete detoxification of waste. But it still leaves behind slags and ashes that amount to some 30 percent of the initial waste volume and that require disposal in a landfill. Finally, separate laws governing air, land, and water may treat the same chemical quite differently. Many compounds strictly regulated under the U.S. hazardous waste law, for example, can legally be released into air or water. Such loopholes create incentives to shift wastes from one environmental medium to another, and may do little to reduce people's exposure to toxics.[41]

If better exposure data were available, lack of knowledge about the toxic effects of most chemicals at various doses would still hinder an accurate assessment of health risks. The vast majority of chemicals have not been fully tested for toxicity, which requires animal experiments that can take several years and cost more than $500,000 per chemical. Even when animal data are available, very different risk assessments emerge, depending on the mathematical model to which the data are applied. One Stanford University researcher found that the estimated cancer risk arising from low-level exposure to the pesticide ethylene dibromide varied by a factor of 1 million depending on the model used.[42]

Epidemiology—the study of the incidence of disease within a population—offers a second approach to assessing health risks. In several ways, however, toxic chemicals pose an epidemiological nightmare. Attributing a delayed health effect to a specific exposure typically requires that the effect have some clinically unique manifestation. But this rarely happens. A cancer induced by a toxic chemical will usually be indistinguishable from a cancer caused by other means. In addition, people usually are exposed to several contaminants simultaneously, whether through drinking water, air, or food. Multiple exposures greatly frustrate efforts to ferret out cause-and-effect relationships. Finally, as Massachusetts Institute of Technology biochemist Dale Hattis writes with Harvard's David Kennedy, "epidemiological studies are notoriously insensitive in detecting health effects from relatively low levels of exposure." Yet most widespread exposures involve contaminants at low concentrations.[43]

Despite these drawbacks, recent epidemiological work among more highly exposed population groups gives cause for concern. Researchers assessing potential health problems in children living near New

> **"Scientists reported a sixfold increase in the risk of non-Hodgkin's lymphoma among Kansas farmers using certain herbicides 20 days or more per year."**

York's Love Canal waste site found seven conditions to be more prevalent than in a control population: seizures, learning problems, hyperactivity, eye irritation, skin rashes, stomach pain, and incontinence. In 1986, scientists reported finding a sixfold increase in the risk of non-Hodgkin's lymphoma, a cancer of the lymphatic system, among Kansas farmers using certain herbicides—especially 2,4-D—for 20 days or more per year. They note that this greater risk among farmers could suggest increased risk to the general population exposed to low levels of these herbicides.[44]

Charles Benbrook, Executive Director of the Board on Agriculture at the National Academy of Sciences, commented on these findings: "For the first time there is clear and rather unequivocal evidence that the environmental exposure to pesticides at low levels causes cancer in man. It has taken a long time for that particular finding to emerge. It involves an old pesticide, one of the most widely used older pesticides, and one that is not really that potent of an oncogen [cause of tumors] if you look at the animal data . . . . I think that the emergence of the new epidemiological data is cause for very serious concern."[45]

Breaking the Pesticide Habit

If chemicals were the only viable way to control crop-damaging insects and weeds, society would have little choice but to live with their associated risks. Fortunately, proven alternatives exist, and others await exploration. A commitment to break agriculture's unhealthy dependence on toxic chemicals is the first step toward realizing the potential of more ecologically sound, economically sustainable pest control methods.

A guiding philosophy known as integrated pest management (IPM) underlies most strategies to reduce pesticide use. IPM recognizes a field of crops as an ecosystem within which many natural forces affecting pests and weeds interact. It draws on biological controls (e.g., natural predators of pests), cultural practices (e.g., planting patterns), genetic manipulations (e.g., pest-resistant crop varieties), and judicious use of chemicals to stabilize crop production while minimizing hazards to health and the environment. The operating goal is not to eradicate insects and weeds but to keep them below the level at which damaging economic losses occur. Under this integrated approach, farmers use chemicals selectively and only when necessary, rather than as the first and primary line of attack.

IPM requires knowledge of a pest's life cycle, behavior, and natural enemies, the way cropping patterns and chemical use affect pest and predator population levels, and many other features of the crop ecosystem. As such, it differs greatly from the packaged variety of pest control provided by today's agrochemical industry. Research and extension are needed to design and implement an effective IPM program. It requires that farmers adopt a new way of thinking about pest management, along with new techniques. But the payoffs can be great. For some farmers, IPM provides an essential escape from the "pesticide treadmill." For most, it offers a welcome response to increasing concerns about chemical costs, health risks from pesticide exposure, and the threat of contaminating their own family's drinking water. As Iowa groundwater specialist George Hallberg recently noted, "The teachable moment is here."[46]

Perhaps no country has worked harder and accomplished more in nonchemical methods of pest control than China. The Chinese began applying the basic tools of IPM long before entomologists had elucidated the concept. For the last three decades, a nationwide pest forecasting system has helped farmers identify, track, and control pest problems. Hundreds of data-collection stations around the country report to their respective provincial forecasting centers, which in turn transmit information on pest populations, the abundance of natural enemies, and weather conditions to some 500 agricultural production units. Between 1979 and 1981, Chinese scientists carried out surveys to locate organisms that could aid farmers in pest control. They identified hundreds of natural control agents for pests of rice, soybeans, tea, and other important crops, creating a rich resource base for expanding biological methods of pest control.[47]

China is well represented among the success stories in IPM and biological control. (See Table 3.) Natural insect enemies keep crop pests in check on some 4.6 million hectares of farmland. In Guangdong Province, a minute wasp that parasitizes the eggs of moth pests

Table 3: Selected Successful Applications of Integrated Pest Management (IPM) and Biological Control (BC)

Country or Region	Crop	Strategy	Effect
Brazil	Soybean	IPM	Pesticide use decreased 80–90 percent over seven years.
Jiangsu Pr., China	Cotton	IPM	Pesticide use decreased 90 percent; pest control costs decreased 84 percent; yields increased.
Orissa, India	Rice	IPM	Insecticide use cut by third to half.
Southern Texas, United States	Cotton	IPM	Insecticide use decreased 88 percent; average net return to farmers increased $77/hectare.
Nicaragua	Cotton	IPM	Early to mid-seventies effort cut insecticide use by a third while yields increased.
Equatorial Africa	Cassava	BC	Parasitic wasp controlling mealybug pest on some 65 million hectares.
Arkansas, United States	Rice/ Soybean	BC	Commercially marketed, fungus-based "bioherbicide" controlling noxious weed.
Guangdong Pr., China	Sugarcane	BC	Parasitic wasp controlling stemborers at one-third the cost of chemical control.
Jilin Pr., China	Corn	BC	Fungus and parasitic wasp providing 80–90 percent control of major corn pest.
Costa Rica	Banana	BC	Pesticide use was stopped; natural enemies reinvaded to control banana pests.
Sri Lanka	Coconut	BC	Parasite found and shipped for $32,250 in early seventies prevents pest damage valued at $11.3 million annually.

Source: Worldwatch Institute, based on various sources.

is controlling sugarcane stemborers at one-third the cost of chemical controls. Cotton growers at Dachiao Commune in Jiangsu Province have implemented an IPM strategy that includes planting sorghum between cotton plants to attract the cotton pests' natural enemies, changing tillage practices, and selectively applying chemicals at the lowest effective doses. Pesticide use dropped 90 percent, pest control costs decreased 84 percent, and cotton yields increased.[48]

China's large pool of agricultural workers and its unique social organization have worked to its advantage in ecologically based pest control. Whether the nation will continue this traditional emphasis under its new market-oriented system of agricultural incentives remains to be seen. In the early eighties, the Ministry of Agriculture helped 10 provinces set up pilot stations for biological control, and the Academy of Agricultural Science has established a special biocontrol institute, evidence of continued commitment. Yet between 1979 and 1984, China's pesticide imports increased more than 250 percent, possibly indicating a shift toward chemical controls.[49]

Over the last decade, Brazil also has advanced impressively toward integrated pest control in its production of soybeans, one of its major export crops. Soybeans are plagued by several insect pests, which variously eat the crops' leaves or suck nutrients out of the pods. With the help of U.S. scientists, an IPM program was developed and tested in trial plots in two of Brazil's leading agricultural states in the mid-seventies. By 1979, EMBRAPA, the national agricultural research institution, had officially adopted the program for soybean production nationwide. Scientists trained extension workers, who then trained farmers in the ideas and techniques of the program. An elaborate promotional campaign was carried out with the help of the communications media, and an "Alert System" was devised to give farmers information and recommendations on pest control.[50]

Brazil's IPM efforts achieved marked success. By the early eighties, about 30 percent of Brazilian soybean growers had adopted IPM. Insecticide use by those growers in 1982 was 80–90 percent less than in 1975, the year before the program began. Yet pesticide use remains heavy in other Brazilian cropping systems, and IPM has not yet benefited the vast majority of small farmers growing food for domes-

> **"Governments seeking to protect water supplies from pesticide contamination and to bolster ailing farm economies can do both simultaneously by investing in IPM."**

tic consumption. Nonetheless, pest control specialist Michael Hansen notes that IPM is now more widely used in soybean production in Brazil than in the United States, and that Brazil's efforts have clearly demonstrated the "practicality and economic value of such a program for a developing country."[51]

Industrial countries also have much to gain from an integrated approach to pest control. Governments seeking to protect water supplies from pesticide contamination and to bolster ailing farm economies can do both simultaneously by investing in IPM. Price-depressing crop surpluses and growing amounts of farm debt make strategies that reduce input costs at least as important as ones that increase yields. Moreover, evidence to date strongly supports IPM as an economical enterprise.

In the United States, IPM's roots date at least to the early twentieth century. By then, scientists were using the principles of applied ecology systematically to design multitactical pest control strategies. With the publication of Rachel Carson's *Silent Spring* in 1962 and the resulting public outcry over the environmental side-effects of uncontrolled pesticide use, the stage was set for IPM to blossom anew. Funding from EPA, the National Science Foundation, and USDA launched the "Huffaker Project" in 1972, a six-year endeavor that greatly advanced the basic tools and methods of IPM. About the same time, the USDA Extension Service undertook 39 pilot projects, which included field demonstrations to educate farmers in IPM principles and techniques.[52]

As of 1984, IPM programs supervised by U.S. extension personnel were under way for nearly 40 crops and collectively covered 11 million hectares, about 8 percent of the nation's harvested cropland area. An evaluation of extension IPM programs published in 1987 showed clearly that farmers adopting IPM strategies have benefited economically. Based on survey results on nine commodities from 15 different states, and considering practices on only one crop per state, farmers using IPM collectively earned $579 million more in profits than they would have otherwise. (See Table 4.) Texas cotton farmers using IPM had net returns per hectare averaging $282 higher than other cotton farmers. Apple growers using IPM in New York and almond growers

Table 4: Estimated Average Annual Economic Benefits from Use of IPM, Selected Cases, United States, Early 1980s

State	Crop	Increase in Net Returns to IPM Users: Farm-level	Statewide
		(dollars per hectare)	(thousand dollars)
California	almonds	769	96,580
Georgia	peanuts	154	62,600
Indiana	corn	72	134,230
Kentucky	stored grain	< 1	890
Massachusetts	apples	222	400
Mississippi	cotton	122	29,680
New York	apples	528	33,000
North Carolina	tobacco	6	780
Northwest[1]	alfalfa seed	132	2,420
Texas	cotton	282	215,830
Virginia	soybeans	10	2,570
Total			578,980

[1]Idaho, Nevada, Montana, Oregon, and Washington.

Source: Virginia Cooperative Extension Service, Virginia Tech and Virginia State Universities, in cooperation with USDA-Extension Service, *The National Evaluation of Extension's Integrated Pest Management (IPM) Programs* (Washington, D.C.: USDA, 1987).

using IPM in California had per-hectare profits $528 and $769 greater, respectively, than nonusers.[53]

IPM programs specifically aimed at reducing pesticide use, rather than just at increasing profits, have achieved some impressive results. U.S. IPM efforts, for example, were intensified in the early seventies on cotton, grain sorghum, and peanuts. By 1982, insecticide applications on these crops had dropped dramatically. Insecticide use per hectare of planted sorghum decreased 41 percent; of cotton, 75 percent; and of peanuts, 81 percent. (See Table 5.) In contrast, insecticide use on the areas planted in corn and soybeans—crops that received minimal IPM—slightly increased. As a result, corn replaced cotton as the crop receiving the greatest share of insecticides in the United States.[54]

Despite advances, IPM's potential has barely been tapped. Targeting IPM efforts to areas of heavy pesticide use and strong potential for groundwater contamination, for example, could lessen risks from toxic chemicals in drinking water. One such area is the U.S. Corn Belt. Farmers typically combat corn rootworms by applying insecticides to the soil; in sensitive hydrologic settings, these chemicals can easily leach into groundwater. Kenneth Ostlie, an IPM specialist at the University of Minnesota, points out that farmers in his state have shown that a cultural technique—crop rotation—is an effective way to reduce this risk. A decade ago, farmers began rotating corn with other crops instead of growing it continuously, in order to reduce fertilizer, fuel, and other input costs. As an added bonus, pest problems diminished with crop rotation, reducing the need for pesticides. Over the last decade, soil insecticide use on Minnesota corn acreage has dropped 45 percent.[55]

Table 5: Effects of IPM on Insecticide Use, United States, 1971-82

		Insecticide Use		
Crop	**Use of IPM**	**1971**	**1982**	**Change**
		(kilograms/hectare)		(percent)
Corn	minimal	0.38	0.41	+ 8
Soybeans	minimal	0.15	0.17	+13
Grain Sorghum	intensive	0.30	0.18	−41
Cotton	intensive	6.63	1.68	−75
Peanuts	intensive	4.48	0.86	−81

Source: R.E. Frisbie and P.L. Adkisson, "IPM: Definitions and Current Status in U.S. Agriculture," in Marjorie A. Hoy and Donald C. Herzog, eds., *Biological Control in Agricultural IPM Systems* (Orlando, Fla: Academic Press, Inc., 1985).

Similarly, an IPM program on Long Island that includes shifting away from monoculture potato farming may be necessary both to keep potato farmers in business and to protect the region's drinking water from pesticide contamination. Studies suggest that crop rotation could eliminate the need for two to four insecticide sprayings per season. Having crop rotation work as a pest control strategy, however, usually requires that most if not all farmers in a pest-plagued region adopt it. Otherwise, pests will invade fields under rotation from neighboring lands under continuous cultivation. Achieving such regional coordination is a major challenge for some IPM strategies.[56]

Biological methods of pest control, either alone or as part of an IPM design, can provide some of the most elegant and long-lasting solutions to pest problems. In "classical" biological control, a beneficial organism is introduced into a pest-plagued area and, it is hoped, becomes a permanent part of the agroecosystem. The pest and the introduced natural enemy reach a population balance that keeps pest damage below the economic threshold. Classical biocontrol thrived in the United States during the late nineteenth and early twentieth centuries. Since the 1860s, scientists have introduced some 300 organisms worldwide in classical control programs.[57]

Finding, testing, and introducing a biological agent is time-consuming and costly. But in contrast to control by chemicals, which requires a large annual investment, successful classical biocontrol requires minimal yearly costs, mainly those associated with protecting and conserving the introduced species. Benefits accrue for many years from the initial investment. For example, about $750,000 was spent during the forties on locating and introducing a European beetle to control a toxic range weed in California. To date, accumulated savings from this effort are estimated at more than $100 million.[58]

Among the most exciting biological control efforts now under way is that to protect African cassava from the ravaging effects of mealybugs and green spider mites. The spider mite was first detected in Uganda in 1971, and the mealybug in Zaire in 1973. With no apparent natural enemies to keep them in check, both pests spread rapidly. By 1982, the mealybug had infiltrated a large portion of the cassava belt, which

"A tiny wasp called *Epidinocarsis lopezi* now effectively controls the mealybug over 65 million hectares in 13 countries of Africa's cassava belt."

spans 34 countries—from Senegal in the west across to Kenya in the east, and south to Angola and southern Mozambique. Some 200 million Africans rely on cassava as a major food crop, and the vast majority of it is grown on small plots by subsistence farmers. Together the two pests slashed cassava yields by 10–60 percent, causing losses estimated at $2 billion annually.[59]

Officials ruled out a massive pesticide program because the infrastructure was lacking to deliver chemicals to the subsistence farmers in need. Instead, the Nigeria-based International Institute of Tropical Agriculture (IITA) launched a major biological control effort with funding from several foreign governments and technical support from the U.K.-based Commonwealth Institute for Biological Control. Extensive searching in Latin America, cassava's place of origin, turned up some 30 species of natural enemies of the mealybug. Several were quarantined and subsequently released in Africa.[60]

To date, a tiny wasp called *Epidinocarsis lopezi*, which parasitizes the mealybugs' eggs, has produced remarkable results. *E. lopezi* now effectively controls the mealybug over 65 million hectares in 13 countries of the cassava belt. African farmers are again growing cassava where mealybug damage had previously decimated the crop. Scientists plan to introduce the wasp into the entire pest-infested area; whether it will prove effective in different ecological settings remains to be seen. Meanwhile, IITA hopes to establish national biocontrol programs and laboratory facilities in several countries, to train more African scientists in pest control methods, and to mass-produce more wasps. The cassava mealybug effort so far has cost about $12 million, less than half the current cost of commercializing one chemical pesticide. Assuming *E. lopezi* provides the permanent control a successful introduction should, the annual benefits to African farmers will far exceed the project's costs.[61]

Another type of biocontrol strategy involves releasing large numbers of a pest's natural enemy during critical periods of the growing season to temporarily suppress the pest population, much the way chemical pesticides do. Probably the most widely used control agent of this type is *Trichogramma*, a tiny wasp that parasitizes the eggs of certain butterflies and moths, preventing them from developing into crop-damaging caterpillars. Useful in both temperate and tropical regions, *Trichogramma* now controls moth pests on an estimated 17 million hectares of cropland worldwide.[62]

The Chinese have long relied on *Trichogramma*, and continue to release large numbers of them to control pests of corn, cotton, rice, sugarcane, and other crops. Farmers in El Salvador were annually releasing *Trichogramma* on about 10,000 hectares of cotton during the late seventies, eliminating the need for 10 pesticide applications per growing season. In India, Biocontrol Research Laboratories, the nation's first commercial insectary, supplies *Trichogramma* for use in controlling a variety of sugarcane and cotton pests. They come on a rectangular "Tricho Card"—20,000 wasps per card—along with directions for their release. The company recommends a minimum dosage of 40,000 wasps (two cards' worth) per acre, which according to its price list would cost 36 rupees ($2.75).[63]

So far, most IPM and biological control strategies have been directed at insect pests, and thus have reduced insecticide use. Of growing importance, however, are nonchemical methods of controlling weeds, the other major class of crop-damaging pests. Chemical weedkillers claim a growing share of the pesticide market in many countries, and they are turning up frequently in rural groundwater supplies. Though biological methods of weed control date back more than a century, only recently have they begun to show potential for commercial field and orchard crops, where most herbicide applications occur.[64]

The use of fungi, bacteria, and other disease-causing agents as "bioherbicides" shows perhaps the greatest near-term promise for supplanting chemicals in weed control. Developing a bioherbicide involves locating diseased members of the targeted weed population, isolating the pathogen—the organism causing the disease—from the weed's tissue, and mass-producing the pathogen for distribution. Good candidates for commercialization will effectively control the weed without attacking crops and other desirable plants. Biological herbicides are applied in much the same way as their chemical counterparts, but largely eliminate health risks to farm workers and the threat of contaminating soils, food, and drinking water.[65]

In recent years, two bioherbicides relying on a fungus as the working agent have entered the U.S. market. DeVine, produced by Abbott Laboratories, controls the milkweed vine in Florida citrus groves. Also known as strangler vine, the weed climbs the tree's trunk, covers the crown, and smothers the tree. Groves treated with DeVine in 1980 apparently were still protected from damage five years later. Collego, marketed by the Upjohn Company, has achieved 90 percent control of northern jointvetch, a troublesome weed in Arkansas rice and soybean fields. Just 13 months after Upjohn submitted toxicology and other testing data to EPA, and two years after the bioherbicide's creation, Collego became fully registered under U.S. pesticide law—a remarkably rapid move from product development to commercial use.[66]

Altering cultural practices and cropping patterns can also help control weeds, just as it does insect pests. Research efforts under way include intercropping—for example, growing a nitrogen-fixing legume between rows of wheat. The legume competes with weeds, keeping them in check, besides adding nitrogen to the soil for the next season's crop. Planting cover crops that inhibit the germination or growth of weeds also shows promise. Such cropping patterns make use of a phenomenon known as allelopathy, the inhibition of one plant by another through the release of natural toxins. Researchers at Michigan State University have found, for example, that leaving residues of rye, sorghum, wheat, or barley on a field can provide up to 95 percent weed control for a month or two.[67]

Such methods show promise of lessening one of the negative aspects of conservation tillage. By relying more on chemicals and less on mechanical tillage to control weeds, conservation tillage has presented an uneasy trade-off between reduced soil erosion and increased risks of groundwater contamination. But according to agricultural writer Charles E. Little, with new biological weed controls, new rotations with allelopathic cover crops, new chemicals effective at very low doses, and more efficient ways of applying chemicals, "a new phase in the ecology of conservation tillage may well be about to begin, with greatly reduced environmental impacts."[68]

The future of IPM, biological control, and other pesticide reduction practices, while promising, is clouded by several factors. New techniques in biotechnology could serve either to promote or to undermine nonchemical methods of pest management. Scientists can use gene-splicing methods, for example, to build pest resistance into crop varieties better and more quickly. Crops less damaged by pests and disease would need fewer chemicals applied to them.

On the other hand, some two dozen chemical and biotechnology companies are researching ways to make crops resistant to chemical herbicides. Some herbicides discriminate poorly between crops and weeds, so farmers can only use them prior to planting or in very selective locations. Crops engineered to resist weed-killers could pave the way for broader use of chemicals. American Cyanamid, for example, has teamed up with Pioneer Hi-Bred International, a seed company, to breed corn varieties resistant to a new line of herbicides. Such input packages promoted by the agrochemical industry could lead to sustained reliance on chemicals at the expense of alternative pest control methods.[69]

In recent years, several U.S. chemical companies have developed herbicides that apparently pose little risk to people, fish, and other animals, and that either break down rapidly in the environment or will not leach into groundwater. These seemingly safer chemicals could also undermine biological and other nonchemical techniques by offering a new and tempting quick fix. Indeed, the vast majority of corporate R&D in herbicide resistance is directed toward these newer compounds, which could bring them into more widespread use among farmers who can afford them.[70] Given the number of adverse chemical effects that have taken society by surprise, it seems unwise to place faith and resources solely in these new products. Safer chemicals are certainly a welcome development, but they will best serve farmers and society if their use is integrated with other promising pest control methods.

Rethinking Industrial Waste Management

How to handle society's toxic chemical waste now ranks among the top environmental issues in most industrial countries. As mentioned earlier, in October 1986 the United States enacted legislation calling for an $8.5-billion Superfund for the cleanup of old waste sites. Indeed, for many countries taking on the challenge, locating and cleaning up all the leaking landfills and waste lagoons scattered across the industrial landscape will be among the highest priced items on their environmental agendas.

Unfortunately, remedying the legacies of past mismanagement only begins to address the toxics dilemma. Unless the wastes currently produced are better managed, new threats will simply replace the old ones, committing society to a costly and perpetual mission of toxic chemical cleanups. A number of countries offer valuable lessons in improving industrial waste management. Few if any nations, however, have adequately tackled the third major task: controlling the total volume of waste generated. Without concerted efforts to reduce, recycle, and reuse more industrial waste, the quantities produced will overwhelm even the best treatment and disposal systems, and the goal of risk-minimizing, sustainable waste management will remain elusive.

Most countries still rely predominantly on land disposal methods—such as landfills, lagoons, and injection wells—for their hazardous wastes. But in several parts of Europe advanced technologies and effective institutional arrangements have combined to create management systems that appear to prevent most waste from being released into the environment in hazardous forms. Two such programs with comparatively long track records are those in Denmark and the West German state of Bavaria.[71]

Though Denmark is not heavily industrialized, the risks of hazardous waste disposal became a major concern there about 1970, much earlier than in most other countries. Relying almost exclusively on groundwater for their drinking water supplies, the Danes had ample reason for that concern. They established a system, now in operation for more than a decade, that treats, detoxifies, or destroys most of the nation's hazardous waste.[72]

At the heart of the Danish system is Kommunekemi, a treatment facility located in the town of Nyborg on the island of Fyn, the nation's geographic center. Industries deliver their toxic waste to the nearest of 21 transfer stations scattered throughout the country. Some 300 smaller collection units—at least one per municipality—accept paints, solvents, and other hazardous wastes from households. That waste also goes to one of the transfer stations, from which all waste gets transported to Kommunekemi.[73]

Only about a quarter of the total waste volume entering the facility ultimately gets landfilled. Almost all of it consists of relatively nontoxic and immobilized residues from the various treatment processes. Rotary kiln incinerators operating at 1200° Celsius detoxify solvents, oily sludge, and organic chemicals. Waste oils, after treatment, help fuel the barrel-shaped kilns. Steam from the incinerators feeds into Nyborg's district heating system, supplying nearly half of the town's heating energy. Electroplating and other inorganic chemical wastes are chemically and physically treated to detoxify the most hazardous compounds and to filter out heavy metals. The resulting filter cakes are landfilled apart from other wastes so that the metals can later be reclaimed. Water from the inorganic treatment plant goes to the municipal sewer system, and flue gases from the incinerators enter the atmosphere. According to plant personnel, releases of dioxin are not a problem at Kommunekemi.[74]

In West Germany, the länder each operate their own waste systems. In Bavaria, as in Denmark, integrated treatment facilities equipped with incinerators, inorganic chemical treatment plants, and secure landfills form the technological backbone of hazardous waste management. A network of collection stations feeds wastes into Bavaria's regional facilities. Prices charged by the primary management company vary from 70 German marks ($38) per ton for relatively nontoxic wastes to 680 marks ($368) per ton for highly toxic pesticides, comparatively low figures that reflect government subsidies for plant construction. In a small town called Geretsried, south of Munich, a recycling facility accepts between 2,000 and 3,000 tons of spent solvents each year, reclaiming about half that quantity for resale.[75]

A critical but controversial feature of the Danish and Bavarian systems is the creation of a monopoly market for a publicly controlled company established to operate the waste system. With limited ex-

"Industries in Denmark deliver their toxic waste to the nearest of 21 transfer stations scattered throughout the country."

ceptions, Danish industries are required to send their wastes to Kommunekemi a/s, a corporation formed and owned by Danish municipalities. Similarly, in Bavaria, Gesellschaft zur Beseitigung von Sondermüll in Bayern mbH (known as GSB) has a monopoly on 70 percent of Bavaria's waste, all except that generated in Middle Franconia, which has its own monopoly company. The Bavarian government owns about 80 percent of GSB, with municipalities and industries splitting ownership of the remaining fifth.[76]

In sharp contrast to this public monopoly approach, the U.S. hazardous waste system is characterized by privately owned, competitive facilities operating with little or no public monies. About 95 percent of U.S. industry's hazardous waste is disposed of on the site where it is generated; scattered commercial facilities handle the rest. Facilities typically offer only one or two methods of treatment or disposal, as opposed to the integrated plants handling all waste types in Bavaria and Denmark. The U.S. government's role is strictly regulatory: It sets construction and operating standards with which the waste management facilities are to comply.[77]

Neither approach will work best in all situations, since any institutional arrangement must mesh with the prevailing ideology and political culture. But experience to date suggests that a publicly controlled monopoly on hazardous waste management offers some advantages over reliance on a private, competitive market. Since one company, rather than dozens or hundreds, manages waste in a given region, the monopoly approach offers greater control and easier enforcement. By obligation, the public companies accept all wastes, whereas a private market might not provide adequately for those that are difficult to treat profitably. Also, as hazardous waste specialists Gary Davis and Bruce Piasecki point out, by requiring industries to deliver most of their chemical wastes to specified facilities where management decisions are then made, "Denmark and West Germany have avoided America's two most paralyzing regulatory battles—deciding what qualifies as a regulated toxic waste, and how these wastes should be managed to ensure public safety and environmental quality."[78]

One obvious advantage of the U.S. approach is the potential for greater efficiency from the presence of competition and a profit motive. A new set of U.S. regulations banning land disposal of certain highly toxic wastes should promote greater use of incineration and advanced treatment methods over the next several years. Yet more than a decade after passage of the U.S. hazardous waste law, a comprehensive, smooth-running system is not in place. From available evidence, it seems that programs with a strong public sector management role—such as Bavaria's and Denmark's—have come closer to the goal of minimizing risks from chemical wastes entering the environment.[79]

In recent years, several other nations have adopted programs with public monopoly components. Finland decided in 1980 to establish a chemical waste system patterned after the Danish experience; its integrated treatment facility began operating in 1984. Sweden has also followed the Danish model, with a plant in Norrtorp starting up in 1983. South Korea is apparently the first developing or newly industrialized country to at least partially embrace the Danish and Bavarian approaches. The government has built a 60-ton-per-day (tpd) centralized plant to serve the Seoul area, expected to have begun operating in July 1987. A second facility, with a 100-tpd capacity, is slated for start-up in Ulsan in September 1987, and a third, 10-tpd plant is planned.[80]

Regardless of the type of management system established—public or private, monopoly or competitive—greater efforts are needed to curb the amounts of waste being generated. Rising costs, scarce treatment and disposal capacity, and public opposition to siting new facilities plague hazardous waste programs virtually everywhere. At Denmark's Kommunekemi facility, for example, the incoming waste volume has increased about 17 percent annually for the last three years. The incinerators are operating at full capacity, and a new one is desperately needed. But local opposition may prevent it from being built in Nyborg, requiring construction elsewhere and delays in bringing it on-line.[81]

In the United States, the number of active, land-based hazardous waste sites has dropped by two-thirds over the last year because of

"Strategies to reduce waste differ markedly from the end-of-pipe treatment to which most industries have grown accustomed."

failure to meet deadlines for complying with operating requirements. The nation's 13 commercial incinerators are operating at 90 percent of capacity; local opposition makes siting new facilities extremely difficult. As demand outpaces supply and as tighter regulations are placed on disposal technologies, waste management costs are rising rapidly. Landfill prices have skyrocketed to $240 per ton, a 16-fold increase since the early seventies. Incineration of organics now costs between $500 and $1,200 per ton. Waste management costs for Du Pont, the nation's largest chemical producer, now exceed $100 million annually. Paul Chubb, vice chairman of Du Pont's Manufacturing Committee, says that "an economical and environmentally acceptable" waste management plan may now "hold the key to the success or failure of many of our businesses."[82]

By not producing waste, industries obviously avoid all the costs and risks of treating, storing, transporting, and disposing of it. Strategies to reduce waste differ markedly from the end-of-pipe treatment to which most industries have grown accustomed. They focus on the production process itself, examining where wastes are generated and exploring how they can be reduced. Simple housekeeping measures, such as segregating wastes so they can more easily be reused, sometimes result in surprisingly large waste reductions. Other options include changing manufacturing processes, using different raw materials, and replacing hazardous products with safer substitutes.[83]

Numerous case studies of individual company efforts collectively attest to waste reduction's feasibility and cost-effectiveness. (See Table 6.) The Minnesota Mining and Manufacturing Company (3M) probably has the longest-standing commitment to waste reduction of any major corporation. Through its "Pollution Prevention Pays" program, launched in 1975, the company claims to have halved its generation of wastes and saved nearly $300 million. In the case of Duphar, a large chemical company in the Netherlands, a new manufacturing process cut the amount of waste generated from the production of one of its pesticides by 95 percent. At a Du Pont paint and finish manufacturing plant in Valencia, Venezuela, a new solvent recovery unit eliminated disposal of solvent wastes, saving the company $200,000 per year.[84]

Table 6: Selected Successful Industrial Waste Reduction Efforts

Company/ Location	Products	Strategy and Effect
Astra Södertälje, Sweden	Pharmaceuticals	Improved in-plant recycling and substitution of water for solvents cut toxic wastes by half.
Borden Chem. California, United States	Resins; adhesives	Altered rinsing and other operating procedures cut organic chemicals in wastewater by 93 percent; sludge disposal costs reduced by $49,000/year.
Cleo Wrap Tennessee, United States	Gift wrapping paper	Substitution of water-based for solvent-based ink virtually eliminated hazardous waste, saving $35,000/year.
Duphar Amsterdam, The Netherlands	Pesticides	New manufacturing process cut toxic waste per unit of one chemical produced from 20 kilograms to 1.

Some of the most appealing strategies stem from examining whether a less hazardous product can replace a hazardous one in performing some particular function. The Indianapolis-based Brulin & Company, for example, has released a new line of nonhazardous, biodegradable industrial cleaners called "Worksafe." In contrast to most cleaners on the market, Worksafe products contain no chlorinated or petroleum-based solvents. The effective ingredients are naturally occurring citrus oils called orange terpenes, which make the products safe to use and far less costly to dispose of. In Austria and West Germany, advocates of "soft chemistry"—a takeoff on the seventies' call for "soft energy" alternatives—point to the widespread substitution of water-based for solvent-based paints as one example of a shift to safer products that needs to be broadened.[85]

Du Pont Barranquilla, Colombia	Pesticides	New equipment to recover chemical used in making a fungicide reclaims materials valued at $50,000 annually; waste discharges were cut 95 percent.
Du Pont Valencia, Venezuela	Paints; finishes	New solvent recovery unit eliminated disposal of solvent wastes, saving $200,000/year.
3M Minnesota, United States	Varied	Companywide, 12-year pollution prevention effort has halved waste generation, yielding total savings of $300 million.
Pioneer Metal Finishing New Jersey, United States	Electroplated metal	New treatment system design cut water use by 96 percent and sludge production by 20 percent; annual net savings of $52,500; investment paid back in three years.

Source: Worldwatch Institute, based on various sources.

As a creative, ongoing endeavor that essentially equates waste with inefficiency, waste reduction represents a new way of thinking. Its success hinges on top management making it a priority, since ideas can spring from all phases of a production process. 3M developed a videotape and brochure explaining the goals of its pollution prevention program to employees, and holds award ceremonies to recognize those who develop innovative projects. USS Chemicals rewards employees who develop waste-cutting ideas with a share of the money thereby saved. As of 1986, the company had distributed $70,000 in rewards for projects saving a total of $500,000.[86]

Despite signs of a shift toward waste reduction, the gains achieved so far represent a small share of their potential. Hazardous waste analysts Joel Hirschhorn and Kirsten Oldenburg point out that waste reduction "is not yet a serious pursuit . . . [it] is at the top of everyone's hierarchy as a theoretical goal, although practical consideration and implementation are postponed to a vague future." Indeed, a study of 29 U.S. organic chemical plants conducted by INFORM, an environmental research group in New York, found that the waste reductions achieved by the companies examined, while impressive, amounted to only a minute fraction of the total waste volume the facilities generated. EPA estimates that expanded use of existing techniques could reduce the total U.S. industrial waste stream by 15–30 percent. As the broad-based 3M program suggests, vigorous efforts could do much more.[87]

Along with reducing waste at its source, recycling and reusing waste can slow the volume of chemicals needing treatment and disposal and help keep toxics out of the environment. Many industries recycle a portion of their wastes internally, and to the extent that these quantities are excluded from statistics, recycling rates can be underestimated. Still, in most countries recycling accounts for only a small fraction of the total volume of waste managed.

In Denmark, for example, just 5 percent of the industrial waste handled off-site is currently recycled, most of it solvents. Engineers at the Danish Technological Institute of Copenhagen in Taastrup estimate that 20–25 percent of Denmark's waste stream could be recycled, even assuming no advances in technology. Most of the gains would come from greater recycling of solvents and metals, including mercury, chromium, and copper. Similarly, EPA found that only 4 percent of the hazardous waste generated in the United States in 1981 was recycled. The Congressional Budget Office estimates that 80 percent of waste solvents and 50 percent of the metals in liquid waste streams could be recovered, compared with EPA's findings of 24 percent and 28 percent, respectively, in 1981.[88]

Japan seems to have advanced the furthest of any major industrial country toward recycling and reusing its industrial waste. Of the estimated 220 million tons of waste generated in 1983, more than half was recycled. (See Table 7.) Incineration, dewatering, and other treatment methods eliminated 31 percent of the waste stream, leaving just 18 percent for final disposal. Since these figures apply to all

Table 7: Industrial Waste Management in Japan, 1983

Waste Disposition	Quantity	Share of Total
	(million tons)	(percent)
Total generated	220.5	100
Recycled and reused	112.7	51
Delivered off-site for reuse	(78.5)	(36)
Reused on-site	(34.2)	(15)
Reduced through treatment and incineration	68.9	31
Disposed of	38.9	18

Source: Clean Japan Center, *Recycling '86: Turning Waste into RESOURCES* (Tokyo: 1986).

industrial waste, not just to those specified as hazardous or toxic, they are not strictly comparable to other national estimates. Nonetheless, Japan's accomplishments are impressive.

A unique cooperative relationship between industry and government lies behind much of Japan's success. Plans developed jointly by private firms and local governments set forth waste management goals, and though they are not legally binding, industry treats them as so. All levels of government encourage recycling. Technical staffs in each prefecture and large city advise industries on waste management. An agency called the Clean Japan Center carries out research and demonstration projects to promote new recycling and waste reduction technologies. Since its creation in 1975, the Center has built 10 model plants, including facilities for recycling oil, mercury, old tires, and plastics. In 1985, the Center operated with a modest budget of 900 million yen ($6 million), with half its funding supplied by the Ministry of International Trade and Industry, 40 percent by Japanese industries, and 10 percent by regional governments.[89]

In Japan, North America, and Western Europe, waste exchanges have succeeded to varying degrees in promoting the recycling and reuse of industrial waste. Exchanges operate on the simple premise that one industry's waste can be another's raw material. Most of them serve as information clearinghouses, publishing catalogs of "waste available" and "waste wanted" listings to inform industries of trading opportunities. A successful trade benefits both buyer and seller, since the former reduces its raw material costs and the latter its treatment and disposal costs.

More than a decade of experience with waste exchanges in Europe suggests that only 5–10 percent of a region's waste might practically be recycled via an exchange. Still, in conjunction with other recycling and waste reduction efforts, exchanges can provide a valuable management alternative. In Japan, exchanges have helped create markets for materials that previously had not been recycled, including sludges, slags, and waste plastics. Sixteen nonprofit exchanges currently operate in North America, and several have experienced healthy growth in recent years. The Northeast Industrial Waste Exchange, based in Syracuse, New York, operates a computerized data base service with listings from five regional exchanges that anyone with a computer, modem, and password can gain access to free of charge. Since listings are continuously updated, the data base eliminates the delays inherent in relying on a quarterly or bimonthly catalog, opening up a spot market that could promote more recycling.[90]

Detoxifying the Environment

Current efforts in integrated pest management and industrial waste reduction only hint at the long-term potential of these two strategies to detoxify the environment. Pesticide use in agriculture could probably be halved and industrial waste cut by at least a third over the next decade. Experience to date suggests that farmers and industries would benefit economically, while threats to public health and the environment would diminish. Yet for society to realize these gains,

> **"Pesticide use in agriculture could probably be halved and industrial waste cut by at least a third over the next decade."**

policies and funding priorities need to actively promote these new methods of production in agriculture and industry, rather than undermining them.

Unraveling the near-total reliance farmers have acquired on chemicals will require much greater efforts from agricultural extension workers and researchers to advance nonchemical methods of controlling insects and weeds. IPM programs can take many forms, and not all of them will substantially reduce a farmer's dependence on pesticides. Many fit the definition of IPM because they incorporate the basic techniques of monitoring pests and setting damage thresholds, but they still rely on chemicals as the primary or sole means of pest control.

In the United States, for example, the USDA Extension Service spent $48 million on IPM between 1973 and 1983. This modest public investment has benefited farmers greatly, but so far apparently has not captured the potential societal benefits of reduced chemical use. As indicated earlier, a recent evaluation found that farmers using IPM in 15 states collectively reaped $579 million a year more than they would have otherwise. Though the exact sources of these increased profits are uncertain, most apparently resulted from higher yields, and thus higher gross returns; only in some cases did farmers experience reduced chemical costs. Broadening the use of crop rotation, intercropping, and biological control requires that extension agents work closely with farmers, offering education, training, and demonstrations that these less familiar techniques work. Yet USDA Extension Service funding for IPM has remained at about $7.5 million per year since 1981, just 2 percent of the agency's total budget.[91]

In addition, greater public commitments to research and development are needed in the areas of biological, cultural, and genetic methods of pest control. The private sector has little incentive to develop strategies relying, for example, on crop rotation or allelopathic cover crops because they involve no marketable product. Similarly, as pest control specialist Michael Hansen points out, "any corporation investing in the search for a classical biological control would rapidly find itself out of business; since the product happily reproduces itself for free, the company is left with nothing to sell."

But the public sector is not adequately filling this R&D gap, despite evidence of the societal benefits. Currently, total direct federal funding in the United States for IPM research amounts to about $20 million annually—less than is needed to commercialize one chemical pesticide, and a mere one-tenth of 1 percent of the $26 billion paid to farmers in crop subsidies in 1986.[92]

Revenues to expand research and extension efforts could come from a very modest tax on pesticide sales. Just a 2 percent tax on sales in the United States, which in 1985 totaled nearly $6.6 billion, would yield revenues sufficient to increase the annual federal IPM extension budget 17-fold, the research budget more than sixfold, or the combined research and extension budgets nearly fivefold.[93] Such an increase in resources could go a long way toward meeting a target of halving pesticide use. Moreover, those who paid the tax would likely benefit in the long run from the results of research and the increased extension services carried out with the funds.

In developing countries, IPM and biological control offer promises of reduced poisonings and deaths from toxic pesticides, while simultaneously creating more sustainable crop production systems. Among the institutions promoting alternatives to chemical pesticides for Third World farmers are the International Institute of Tropical Agriculture in Ibadan, Nigeria, which coordinates the Africa-wide Biological Control Project; the U.K.-based Commonwealth Institute of Biological Control (CIBC); the International Rice Research Institute in the Philippines; and research agencies in China, India, Kenya, and Thailand. Both the World Bank and the U.S. Agency for International Development, major lenders for Third World agricultural projects, have written policies that support IPM. But actual implementation of these strategies has lagged. As CIBC Director David Greathead points out, alternatives to pesticides are still usually adopted only as a last resort—for example, when chemicals become too expensive or fail to work against resistant pests—rather than as an integral part of agricultural planning and development.[94]

An important first step for much of the Third World is to stop subsidizing chemical pesticides so heavily. Subsidies encourage farmers to apply more chemicals than is economically justified, undermine

"In November 1986, Indonesia's President Suharto banned the use of 57 insecticides on rice, and essentially made IPM national policy."

the development and use of nonchemical methods, and ultimately increase all the risks associated with toxic farm chemicals. In a study of nine developing countries—three each from Africa, Asia, and Latin America—the World Resources Institute of Washington, D.C., found that pesticide subsidies ranged from 19 percent of real retail costs in China to 89 percent in Senegal. The median was 44 percent. Of the nine, only Pakistan had examined the efficacy of its subsidies and, as a result, largely discontinued them in 1980. By phasing subsidies out and devoting the public funds thereby freed to research and extension in IPM and biological control, governments could do much to promote more ecologically sound and sustainable pest control.[95]

Industrial nations share responsibility for the high health and environmental risks of pesticide use in the Third World. Most of them freely export to developing countries chemicals that are banned or severely restricted within their own borders. Requiring exporters to obtain written consent from the importing country prior to shipping restricted chemicals could help curb the dumping of dangerous pesticides in the Third World. The Netherlands has passed legislation authorizing the government to make prior consent mandatory if a voluntary scheme now being implemented is deemed insufficient. The Organisation for Economic Co-operation and Development, the U.N. Environment Programme, and the U.N. Food and Agriculture Organization each have set guidelines on chemical exports, but none calls for prior consent.[96]

Mounting concerns about pest resistance, chemical costs, groundwater contamination, and health risks have spurred a few governments to act to curb pesticide use. By far the most notable initiative is that of Indonesia. Having achieved self-sufficiency in rice in 1984, the nation now finds that position threatened by the brown planthopper, an insect pest that has acquired resistance to every major rice pesticide. In November 1986, President Suharto banned the use of 57 insecticides on rice, and essentially made IPM national policy. The government planned to train 200,000 farmers in IPM techniques by the summer of 1987. No other country has so strongly supported IPM at such a high official level. Indonesia's effort may set an example other nations trapped in the pesticide treadmill could follow.[97]

In Sweden, a program adopted in 1987 aims at cutting risks from pesticides by half over the next five years. The government has allotted 54 million kronor ($8.4 million) to the effort, which will include two-day courses for the nation's 20,000 farmers, promotion of more efficient chemical spraying equipment, incentives for the use of safer chemicals, and guidance on IPM techniques. A new Danish program aims at cutting pesticide use by 25 percent by 1990, and by a further 25 percent by 1997. The government has imposed a 3-percent tax on pesticides to help pay for increased R&D and educational efforts in nonchemical pest control.[98]

In the United States, Vermont's Governor Madeline Kunin called in May 1986 for the development of plans to curb the government's use of pesticides and for the Agriculture Commissioner to expand efforts promoting pesticide alternatives among farmers. Her policy statement sets no targets, however, and its ultimate effect remains uncertain. New legislation in Nebraska authorizes the state environment department to require specific management practices—including reduced pesticide use—in designated areas where agriculture has contaminated groundwater. And in Iowa, comprehensive groundwater legislation passed in May 1987 includes funding for research and education in improved agricultural practices, among them a lowering of pesticide use. Partial funding will come from new fees imposed on pesticide retailers.[99]

Despite some promising initiatives, IPM will not gain widespread use among U.S. farmers until federal agricultural policies cease to indirectly promote pesticide use. Government programs offer farmers a guaranteed price for certain crops, and, to control crop surpluses, encourage them to idle a portion of their cropland. USDA economist Katherine Reichelderfer points out that the combination leads farmers to maximize yield—and thus guaranteed income—on the land kept in production. They do so with greater use of agricultural inputs, including pesticides, which partially or wholly offsets the pesticide reductions that result from idling land.[100]

By tying conservation priorities to farm programs, the 1985 Food Security Act offered a unique opportunity to redress some of the negative consequences of agricultural practices. Arguably one of the strongest conservation initiatives in decades, the act created a "conservation reserve," which by 1990 will include between 16 million and 18 million hectares of highly erodible cropland. Farmers will receive compensation for planting this land in grass or trees instead of crops, helping to conserve soils and simultaneously to reduce crop surpluses. Already some 8 million hectares have come out of production. Legislation introduced in the Senate in July 1987 would broaden the use of the reserve by making eligible for it cropland associated with the contamination of water supplies. By linking federal efforts to curb crop surpluses with state and local efforts to protect drinking water, such action would increase the societal benefits of the government's multibillion-dollar farm programs.[101]

Regarding industrial chemical wastes, virtually no country has yet designed an effective, long-term strategy. Although some nations, notably several in Western Europe, have established impressive waste management systems, efforts to curb the quantities generated need far greater attention. Spotty data on the types and volumes of toxic waste that industries currently produce make it difficult to set goals. But more vigorous research and development in waste-reducing technologies, technical and financial support to encourage investments in such technologies, and, in some cases, a tax on waste generated could probably cut problem wastes by at least a third in most countries over the next decade.

Several West European nations now show strong commitments to promoting "cleaner technologies" and other methods of curbing toxic pollution. The French government, for example, pays up to half the costs for research into widely applicable waste-minimizing technologies, and offers investment subsidies of 10 percent for demonstrations of pollution prevention techniques. Since 1979, some 86 research efforts have received partial financing and more than 80 demonstration projects have benefited from the investment subsidies. Officials estimate that in 1984, government expenditures to promote cleaner technologies totaled 192 million francs ($31 million), inducing investments by private industry of several times that amount.[102]

The Netherlands, too, has embraced waste reduction as a way of boosting industrial competitiveness while simultaneously protecting the environment. A special Committee on Environment and Industry has aided some 200 clean technology research, development, and demonstration projects, including some on the potential role of biotechnologies in curbing waste production. The Dutch government spends about $8 million per year on such efforts, a large sum for a country of only 14.5 million people.[103]

Both Denmark and West Germany, widely lauded for their waste management systems, have recently stepped up their waste reduction efforts. In February 1987, the Danish Parliament allocated 50 million kroner ($7.1 million) for a three-year pilot program of direct investment subsidies for clean technologies. If the subsidies prove effective in the industries studied, a larger program could be initiated. In West Germany, a 1986 law tightening treatment and disposal requirements is expected to at least double and possibly triple waste management costs, providing a strong economic incentive for waste reduction. In addition, the Ministry for Research and Technology and the federal environmental agency help fund waste minimization R&D projects. A program supporting pilot projects also began in 1986, with funding of 15 million marks ($8.1 million) per year for two years.[104]

Efforts in the United States pale in comparison to these European initiatives. In passing the 1984 hazardous waste law, Congress declared it to be "national policy" that "wherever feasible, the generation of hazardous waste is to be reduced or eliminated as expeditiously as possible." Yet EPA's 1988 budget request for waste minimization activities totals just $398,000—0.03 percent of its $1.5-billion operating program budget, and less than was spent in 1986. In a 1986 report to Congress on waste minimization, EPA proposed only to gather information on industry's activities and to provide assistance to states developing their own programs; the level and kind of assistance remained unclear. Believing no further action should be taken until the effects of the legislated land disposal bans could be judged, the agency recommended it report back to Congress in December 1990, and perhaps then consider additional measures.[105]

"It seems foolhardy to ignore the timeworn truism that an ounce of prevention is worth a pound of cure."

With the United States already facing costs of from $20 billion to $100 billion to clean up old toxic waste sites, it seems especially foolhardy to ignore the timeworn truism that an ounce of prevention is worth a pound of cure. By investing modestly in waste reduction now, the government can avoid future problems and costs arising from waste mismanagement, shortfalls in treatment capacity, and public opposition to siting new facilities.

Legislation introduced in the U.S. Congress in June 1987 contains most of the elements needed for a successful national effort. It would create an Office of Waste Reduction within EPA, giving waste reduction a high-level institutional home and signaling a strong federal commitment. It authorizes up to $18 million for waste reduction activities: $8 million to operate the new office and $10 million for grants to the states. This would provide useful seed money, but greater funding for state efforts seems needed to get the waste reduction ball rolling effectively. OTA analysts point out that spending $200 million over five years on state waste reduction grants could save industry billions of dollars in avoided management costs. Moreover, tax revenues from increased company profits likely would exceed the federal cost of the program.[106] If reallocating a small share of EPA's budget to pay for the program seemed infeasible, a minimal tax on waste—less than $1 per ton—would generate more than enough to launch a strong waste reduction initiative.

A handful of states already have their own programs, but they could benefit greatly from an infusion of federal funds. North Carolina's widely praised Pollution Prevention Pays program, budgeted at $600,000 per year, includes technical assistance, research and education activities, and the provision of matching grants to help businesses and communities implement waste reduction and recycling projects. It focuses, uniquely, not just on waste legally defined as hazardous, but on all pollutants discharged to air, soil, or water. Other states actively promoting waste reduction include California (with funding largely from a waste tax), Illinois, Minnesota, New York, Pennsylvania, Tennessee, and Wisconsin.[107]

Few developing countries have even established the basic foundation of a hazardous waste management system. Most have no regulations governing toxic waste and no facilities capable of adequately treating and disposing of such materials. South Korea appears to be one exception, having fairly comprehensive legislation and two advanced treatment facilities slated to begin operation in 1987. Thailand, too, is taking some positive steps. The government is building a central treatment plant to handle the toxic waste streams from some 200 electroplating shops in the Bangkok area. Also, the Thailand Development Research Institute has investigated pollution prevention technologies in three important manufacturing industries—pulp and paper, textiles, and metal finishing—and researchers believe some will be adopted.[108]

An active exchange of information and experience between governments and industries in industrial countries with policymakers in developing countries could do much to advance the Third World's management of toxic chemicals. On a small scale, such efforts are under way. With funding from the Danish Foreign Ministry, for example, the Danish company ChemControl helped the Indonesian government assess its waste problem and begin to design some solutions. In a program announced in August 1986, three U.S. corporations—Dow Chemical, Exxon, and Mobil—will help train Indonesian environmental officials in industrial environmental management techniques, including hazardous waste management. The program was requested by Indonesia's Minister of Population and Environment, established by the New York-based World Environment Center, and will be funded mostly by the participating companies.[109]

Making industries assume responsibility for more of the societal costs and risks associated with hazardous substances is crucial to fostering a transition to safer chemicals and products. Government regulators often bear the burden of showing that a substance causes unacceptable harm before they can act to restrict or ban it. If, instead, industries had to prove suspect substances safe, and if they faced strict liability for damages caused from the manufacture, use, and disposal of their products, risks would diminish throughout the chemical cycle. Risky substances would be weeded out in industrial laboratories, rather than by a regulatory agency after many years of use.[110]

Voters in California overwhelmingly approved a referendum in 1986 that shifts at least some responsibility for chemical safety over to industry. It prohibits industries from releasing chemicals on a state list of those believed to cause cancer or birth defects in a manner that might allow them to enter drinking water. It also requires the labeling of products containing those chemicals, even in trace amounts. In court actions involving exposures to substances covered by the law, industry bears the burden of proving the contested exposure harmless.[111] If rigorously enforced, the new law in California should provide substantial incentive for the manufacture and use of safer chemicals and products.

A unique convergence of public and private interests now makes it a ripe time to promote alternative pest control methods and better management of industrial chemicals. Both farmers dependent on pesticides and generators of hazardous waste face rising costs and risks associated with their practices. Governments face the complex and expensive task of protecting people from contamination caused by agricultural and industrial chemicals. And a justifiably wary public wants assurance that its water, food, and surrounding environment are safe. Technologies and methods to minimize pesticide use and industrial waste tackle each of these concerns. For everyone's benefit, they deserve promoting.

Notes

1. Entomologist quoted is V.B. Wigglesworth, in William W. Lowrance, *Of Acceptable Risk: Science and the Determination of Safety* (Los Altos, Calif.: William Kaufmann, Inc., 1976).

2. In this paper, the terms hazardous, toxic, chemical, and industrial waste are used loosely and interchangeably, in part because of varying national definitions, to connote wastes containing substances potentially threatening to health. This usage does not conform to the official U.S. definition, under which toxic is a subset of hazardous waste, which also includes wastes that are explosive, flammable, or corrosive.

3. For a brief sketch of the organic chemical industry, see David J. Sarokin et al., *Cutting Chemical Wastes* (New York: INFORM, Inc., 1985); James O. Schreck, *Organic Chemistry: Concepts and Applications* (Saint Louis, Mo.: The C.V. Mosby Company, 1975).

4. U.S. International Trade Commission, *Synthetic Organic Chemicals: United States Production and Sales 1985* (Washington, D.C.: U.S. Government Printing Office, 1986); number of chemicals in use from "The Quest for Chemical Safety," *International Register of Potentially Toxic Chemicals Bulletin*, May 1985; number added annually from Michael Shodell, "Risky Business," *Science '85*, October 1985.

5. Michael J. Dover, *A Better Mousetrap: Improving Pest Management in Agriculture* (Washington, D.C.: World Resources Institute, 1985).

6. U.S. Environmental Protection Agency (EPA), "Pesticide Industry Sales and Usage: 1985 Market Estimates," Washington, D.C., September 1986; "Major Changes Coming in Pesticide Law," *Agricultural Outlook*, October 1986; 70 percent figure from Herman Delvo, agricultural economist, U.S. Department of Agriculture (USDA), Washington, D.C., private communication, August 6, 1987.

7. USDA, Economic Research Service, *Agricultural Resources: Inputs Situation and Outlook Report* (Washington, D.C.: U.S. Government Printing Office, 1987); U.S. International Trade Commission, *1986 U.S. Industrial Outlook* (Washington D.C.: Department of Commerce, 1986); Environment Canada, *State of the Environment Report for Canada* (Ottawa: Minister of Supply and Services, 1986); Michael Hanthorn and Michael Duffy, "Corn and Soybean Pest Management Practices for Alternative Tillage Strategies," in USDA, Economic Research Service, *Outlook & Situation* (Washington, D.C.: U.S. Government Printing Office, 1983).

8. Winand D.E. Staring, *Pesticides: Data Collection Systems and Supply, Distribution and Use in Selected Countries of the Asia-Pacific Region* (Bangkok: United Nations Economic and Social Commission for Asia and the Pacific, 1984); Indian data from Y.P. Gupta, "Pesticide Misuse in India," *The Ecologist*, Vol. 16, No. 1, 1986.

9. United Nations Food and Agriculture Organization (FAO), *FAO Trade Yearbook 1984* (Rome: 1985).

10. U.S. Congressional Budget Office (CBO), *Hazardous Waste Management: Recent Changes and Policy Alternatives* (Washington, D.C.: U.S. Government Printing Office, 1985).

11. Ibid.; EPA, *Report to Congress: Minimization of Hazardous Waste*, Vol. 1 (Washington, D.C.: 1986).

12. CBO, *Hazardous Waste Management*; Mark Crawford, "Hazardous Waste: Where to Put It?" *Science*, January 9, 1987.

13. Franz Defregger, "Status and Trends on the Management of Industrial Hazardous Waste in the Federal Republic of Germany," *Industry and Environment*, Special Issue, 1983; Roger Milne, "Flash point for Britain's toxic-waste cowboys," *New Scientist*, July 31, 1986; U.K. Hazardous Waste Inspectorate, *Hazardous Waste Management: "Ramshackle & Antediluvian"?* (London: 1986).

14. H. Yakowitz, "Some Background Information Concerning Hazardous Waste Management in Non-OECD Countries," paper prepared for Organisation for Economic Co-operation and Development, Paris, 1985; figure on Danish exports from Kim Christiansen and Torben Hansen, "Resource Recovery of Hazardous Wastes," Technological Institute of Copenhagen, Taastrup, Denmark, October 1986; figure on German exports from West German Ministry for Research and Technology, *The Japanese-German Panel for Research and Development on Environment Protection Technology* (Bonn: 1986), and from Eberhard Meller, Bundesverband Der Deutschen Industrie, Cologne, West Germany, private communication, January 28, 1987.

15. Yakowitz, "Some Background Information"; H. Jeffrey Leonard, "Confronting Industrial Pollution in Rapidly Industrializing Countries: Myths, Pitfalls, and Opportunities," *Ecology Law Quarterly*, Vol. 12, No. 14, 1985.

16. H. Jeffrey Leonard, "Hazardous Wastes: The Crisis Spreads," *National Development*, April 1986.

17. Leonard, "Confronting Industrial Pollution"; Leonard, "Hazardous Wastes."

18. Yakowitz, "Some Background Information"; Centre for Science and Environment, *The State of India's Environment 1984-85* (New Delhi: 1985); "China Plans Curbs on Solid Wastes," *China Daily*, May 1, 1985.

19. M.J. Stoken, "Hazardous Waste Management—Hong Kong's Experience and Proposals," Hong Kong Environmental Protection Agency, 1985; Malaysian Department of the Environment, "Country Report: Malaysia," paper prepared for Workshop on Policies and Strategies for Managing Hazardous Wastes in Asia and the Pacific, Singapore, May 7-9, 1986.

20. National Research Council (NRC), *Toxicity Testing: Strategies to Determine Needs and Priorities* (Washington, D.C.: National Academy Press, 1984); Charles Benbrook, Executive Director, Board on Agriculture, National Academy of Sciences, Washington, D.C., private communication, May 1986.

21. Foo Gaik Sim, *The Pesticide Poisoning Report* (Penang, Malaysia: International Organization of Consumers Unions, 1985); Dover, *A Better Mousetrap*.

22. Gupta, "Pesticide Misuse in India"; Centre for Science and Environment, *State of India's Environment*; Sean L. Swezey et al., "Nicaragua's Revolution in Pesticide Policy," *Environment*, January/February 1986.

23. David Weir and Mark Schapiro, *Circle of Poison: Pesticides and People in a Hungry World* (San Francisco: Institute for Food and Development Policy, 1981); A. Lees et al., "The Effects of Pesticides on Human Health," Minutes of Evidence presented to the Agriculture Committee, House of Commons, London, May 15, 1986; Shelley A. Hearne, *Harvest of Unknowns* (New York: Natural Resources Defense Council, 1984); Irvin Molotsky, "FDA Is Faulted On Imported Food," *New York Times*, December 4, 1986; Lawrie Mott and Martha Broad, *Pesticides in Food* (San Francisco: Natural Resources Defense Council, 1984).

24. NRC, *Regulating Pesticides in Food: The Delaney Paradox* (Washington, D.C.: National Academy Press, 1987).

25. Lees et al., "Effects of Pesticides on Human Health."

26. Office of Ground-Water Protection, "EPA Ground-Water Protection Strategy: FY 1985 Status Report," EPA, Washington, D.C., updated; George R. Hallberg, "From Hoes to Herbicides: Agriculture and Groundwater Qual-

ity," *Journal of Soil and Water Conservation*, November/December 1986; Charles M. Benbrook and Phyllis B. Moses, "Engineering Crops to Resist Herbicides," *Technology Review*, November/December 1986.

27. EPA survey plans from the *Water Newsletter* (Water Information Center, Plainview, N.Y.), November 30, 1986; Elizabeth G. Nielsen and Linda K. Lee, *The Magnitude and Costs of Groundwater Contamination from Agricultural Chemicals: A National Perspective* (Washington, D.C.: USDA, 1987).

28. Nielsen and Lee, *Groundwater Contamination from Agricultural Chemicals*; Donald V. Feliciano, "Agricultural Effects on Groundwater Quality," Congressional Research Service, Library of Congress, Washington, D.C., October 1986.

29. George W. Ware, *Fundamentals of Pesticides* (Fresno, Calif.: Thomson Publications, 1986); David Bull, *A Growing Problem: Pesticides and the Third World Poor* (Oxford: OXFAM, 1982); Perseu Fernando dos Santos, EMBRAPA, Jaguariúna, Brazil, private communication, April 9, 1987; Nielsen and Lee, *Groundwater Contamination from Agricultural Chemicals*.

30. Robert L. Metcalf, "Changing Role of Insecticides in Crop Protection," *Annual Review of Entomology*, Vol. 25, 1980; Michael Dover, "Getting Off the Pesticide Treadmill," *Technology Review*, November/December 1985; W.C. Shaw, "Integrated Weed Management Systems Technology for Pest Management," *Weed Science*, Supplement to Vol. 30, 1982; Michael J. Dover and Brian A. Croft, "Pesticide Resistance and Public Policy," *BioScience*, February 1986.

31. George P. Georghiou, "The Magnitude of the Resistance Problem," in NRC, Board on Agriculture, *Pesticide Resistance: Strategies and Tactics for Management* (Washington, D.C.: National Academy Press, 1986).

32. Perry L. Adkisson et al., "Controlling Cotton's Insect Pests: A New System," *Science*, April 2, 1982; Swezey et al., "Nicaragua's Revolution in Pesticide Policy."

33. Dover and Croft, "Pesticide Resistance and Public Policy"; Patrick W. Holden, *Pesticides and Groundwater Quality: Issues and Problems in Four States* (Washington, D.C.: National Academy Press, 1986).

34. Dover and Croft, "Pesticide Resistance and Public Policy"; Georghiou, "Magnitude of the Resistance Problem."

35. West German Ministry for Research and Technology, *The Japanese-German Panel on Environment Protection Technology*; Kim Christiansen, Technological Institute of Copenhagen, Taastrup, Denmark, private communication, February 2, 1987.

36. CBO, *Hazardous Waste Management*; Office of Drinking Water, "Surface Impoundment Assessment National Report," EPA, Washington, D.C., 1981.

37. RCRA/CERCLA Hotline, EPA, Washington, D.C., July 31, 1987; World Resources Institute/International Institute for Environment and Development (WRI/IIED), "Managing Hazardous Wastes: The Unmet Challenge," in *World Resources 1987* (New York: Basic Books, 1987); U.S. Congress, Office of Technology Assessment (OTA), *Superfund Strategy* (Washington D.C.: U.S. Government Printing Office, 1985); Mark Crawford, "Toxic Waste, Energy Bills Clear Congress," *Science*, October 31, 1986.

38. Veronica I. Pye and Ruth Patrick, "Ground Water Contamination in the United States," *Science*, August 19, 1983; OTA, *Protecting the Nation's Groundwater from Contamination*, Vol. 1 (Washington, D.C.: U.S. Government Printing Office, 1984); "Congress Passes Bill for Renewal of Safe Drinking Water Act with Groundwater Protection Provisions," *The Groundwater Newsletter* (Plainview, N.Y.), May 30, 1986.

39. Gail Porter, "Controlling Pollution in the Great Lakes Basin," *News Report* (NRC, Washington, D.C.), December 1985/January 1986; Ulla-Britta Fallenius, The Swedish Environmental Protection Board, Solna, Sweden, private communication, February 5, 1987.

40. Eugeniusz Pudlis, "Poland: Heavy Metals Pose Serious Health Problems," *Ambio*, Vol. 11, 1982; Jean Pierre Lasota, "Darkness at Noon," *The Sciences*, July/August 1987.

41. Herr Ulrich Materne and Helga Retsch-Preuss, Gesellschaft zur Beseitigung von Sondermüll in Bayern mbH (GSB), Munich, West Germany, private communication, January 22, 1987; WRI/IIED, "Managing Hazardous Waste"; see also OTA, *Technologies and Management Strategies for Hazardous Waste Control* (Washington, D.C.: U.S. Government Printing Office, 1983).

42. NRC, *Toxicity Testing*; Stanford researcher cited in Dale Hattis and David Kennedy, "Assessing Risks from Health Hazards: An Imperfect Science," *Technology Review*, May/June 1986.

43. NRC, *Drinking Water and Health*, Vol. 6 (Washington D.C.: National Academy Press, 1986); Hattis and Kennedy, "Assessing Risks."

44. Beverly Paigen et al., "Prevalence of Health Problems in Children Living Near Love Canal," *Hazardous Waste & Hazardous Materials*, Vol. 2, No. 1, 1985; Aaron Blair et al., "Cancer and Pesticides Among Farmers," in The Freshwater Foundation, *Pesticides and Groundwater: A Health Concern for the Midwest* (Navarre, Minn.: 1987).

45. Charles M. Benbrook, Reactor Panel statement, in Freshwater Foundation, *Pesticides and Groundwater: Health Concern for Midwest.*

46. George Hallberg, "Overview of Agricultural Chemicals in Groundwater," talk given at the symposium Ground Water Protection: Potential for Avoiding Contamination by Agricultural Chemicals, Institute for Alternative Agriculture (IAA), Washington, D.C., March 3, 1987.

47. L. Brader, "Integrated Pest Control in the Developing World," *Annual Review of Entomology*, Vol. 24, 1979; Qu Geping, "Biological Control of Pests in China," *Mazingira*, Vol. 7, No. 2, 1983; Marcos Kogan, University of Illinois, Champaigne-Urbana, Ill., private communication, May 20, 1987.

48. Qu, "Biological Control in China"; M.J.W. Cock, "The Use of Parasitoids for Augmentative Biological Control of Pests in the People's Republic of China," *Biocontrol News and Information* (Silwood Park, United Kingdom), September 1985; see also Robert L. Metcalf and Arthur Kelman, "Integrated Pest Management in China," *Environment*, May 1981.

49. Qu, "Biological Control in China"; pesticide imports from FAO, *Trade Yearbook 1984*.

50. Michael Hansen, "Escape from the Pesticide Treadmill: Alternatives to Pesticides in Developing Countries," preliminary report, Institute for Consumer Policy Research, Consumers Union, Mount Vernon, N.Y., 1986.

51. Décio L. Gazzoni and Edilson B. de Oliveira, "Soybean Insect Pest Management in Brazil–II. Program Implementation," in P. Matteson, ed., *Proceedings of the International Workshop in Integrated Pest Control for Grain Legumes* (Brazília: EMBRAPA, 1984); Michael Hansen, Institute for Consumer Policy Research, Consumers Union, Mount Vernon, N.Y., private communication, April 10, 1987; quote from Hansen, "Alternatives to Pesticides in Developing Countries."

52. R.E. Frisbie and P.L. Adkisson, "IPM: Definitions and Current Status in U.S. Agriculture," in Marjorie A. Hoy and Donald C. Herzog, eds., *Biological Control in Agricultural IPM Systems* (Orlando, Fla.: Academic Press, Inc., 1985.)

53. Virginia Cooperative Extension Service, Virginia Tech and Virginia State, and USDA-Extension Service, *The National Evaluation of Extension's Integrated Pest Management (IPM) Programs* (Washington, D.C.: USDA, 1987).

54. Frisbie and Adkisson, "IPM: Definitions and Current Status in U.S. Agriculture."

55. Kenneth Ostlie, "Integrated Pest Management: A Tool To Manage the Risk of Ground Water Contamination," in Freshwater Foundation, *Pesticides and Groundwater: Health Concern for Midwest.*

56. Holden, *Pesticides and Groundwater Quality*; need for regional coordination from Hansen, private communication, July 22, 1987.

57. Suzanne W.T. Batra, "Biological Control in Agroecosystems," *Science,* January 8, 1982.

58. Ibid.; for additional economic assessments of biocontrol projects, see D.J. Greathead and J.K. Waage, *Opportunities for Biological Control of Agricultural Pests in Developing Countries* (Washington, D.C.: World Bank, 1983).

59. International Institute of Tropical Agriculture (IITA), *Root and Tuber Improvement Program: Research Highlights 1981-1984* (Ibadan, Nigeria: 1985); Hansen, "Alternatives to Pesticides in Developing Countries."

60. IITA, *Research Highlights*; Hansen, "Alternatives to Pesticides in Developing Countries."

61. IITA, *Annual Report and Research Highlights 1985* (Ibadan, Nigeria: 1986); Hansen, "Alternatives to Pesticides in Developing Countries"; Jeffrey K. Waage, Research Director, Commonwealth Institute of Biological Control, Silwood Park, United Kingdom, private communication, January 20, 1987.

62. Greathead and Waage, *Opportunities for Biological Control in Developing Countries.*

63. Brader, "Integrated Pest Control in the Developing World"; Qu, "Biological Control in China"; Biocontrol Research Laboratories, price list, Bangalore, Karnataka, India, undated.

64. Sara S. Rosenthal et al., *Biological Methods of Weed Control* (Fresno, Calif.: Thomson Publications, 1984).

65. P.C. Quimby, Jr., and H.L. Walker, "Pathogens as Mechanisms for Integrated Weed Management," *Weed Science*, Supplement to Vol. 30, 1982; Rosenthal, *Biological Methods of Weed Control*.

66. Donald S. Kenney, "DeVine—The Way It Was Developed—An Industrialist's View," and R.C. Bowers, "Commercialization of Collego—An Industrialist's View," *Weed Science*, Vol. 34, Supplement 1, 1986; Rosenthal, *Biological Methods of Weed Control*.

67. R.J. Aldrich, *Weed-Crop Ecology: Principles in Weed Management* (North Scituate, Mass.: Breton Publishers, 1984); Alan R. Putnam et al., "Exploitation of Allelopathy for Weed Control in Annual and Perennial Cropping Systems," *Journal of Chemical Ecology*, May 1983.

68. Maureen K. Hinkle, "Problems with Conservation Tillage," *Journal of Soil and Water Conservation*, May/June 1983; Charles E. Little, "The No-Till Revolution," *Country Journal*, September 1985; Charles E. Little, "Beyond the Mongongo Tree: Good News about Conservation Tillage and the Environmental Tradeoff," *Journal of Soil and Water Conservation*, January/February 1987.

69. Benbrook and Moses, "Engineering Crops to Resist Herbicides."

70. Ibid.

71. Author's meetings with various officials and hazardous waste specialists in several European countries, January-February 1987; see also Bruce Piasecki and Gary A. Davis, *America's Hazardous Waste Management Future: Lessons from Europe* (Westport, Conn.: Greenwood Press, in press).

72. For thorough descriptions of the Danish system, see Kenneth Geiser et al., *Foreign Practices in Hazardous Waste Minimization* (Medford, Mass.: Tufts University Center for Environmental Management, 1986), and Bruce Piasecki and Gary Davis, "A Grand Tour of Europe's Hazardous-Waste Facilities," *Technology Review*, July 24, 1984.

73. Mogens Palmark, "The Hazardous Waste Management System in Denmark," paper presented at the Symposium on Pollution Control, Kuala Lumpur, Malaysia, October 29-31, 1985.

74. Author's tour of plant, January 30, 1987; Kommunekemi a/s, "Kommunekemi," Nyborg, Denmark, undated; Palmark, "Hazardous Waste Man-

agement in Denmark"; Per Riemann, Kommunekemi a/s, Nyborg, Denmark, private communication, January 30, 1987.

75. GSB, "Kostenliste" (Price List), Munich, West Germany, January 1, 1987; Materne and Retsch-Preuss, private communication; Klaus Schneider, Chief of Geretsried solvent recycling facility, GSB, Geretsried, West Germany, private communication, January 23, 1987.

76. Riemann, private communication; GSB, "Disposal of Special Refuse in Bavaria," Munich, West Germany, October 1983; Materne and Retsch-Preuss, private communication.

77. Crawford, "Hazardous Waste: Where to Put It?"; Steve R. Drew, Regional Community Relations Manager, Chemical Waste Management, Inc., Newark, Calif., private communication, May 4, 1987.

78. For a detailed discussion of the public monopoly versus private competitive approaches, see Piasecki and Davis, *Lessons from Europe*; quote from Piasecki and Davis, "Grand Tour of Europe's Hazardous-Waste Facilities."

79. Piasecki and Davis, *Lessons from Europe*; Rochelle L. Stanfield, "Drowning in Waste," *National Journal*, May 10, 1986.

80. Chemcontrol a/s, "3rd International Symposium on Operating European Hazardous Waste Management Facilities—Final Program," Odense, Denmark, September 16-19, 1986; George Garland, "Report on Consultancy to the Republic of Korea for the World Health Organization," April 16, 1987, provided by Garland, Office of Solid Waste, EPA, Washington, D.C.

81. Riemann, private communication.

82. Mark Crawford, "OTA Urges Waste Reduction as Dump Sites Close," *Science*, September 26, 1986; Crawford, "Hazardous Waste: Where to Put It?"; cost figures from EPA, *Report to Congress: Minimization of Hazardous Waste*; Paul A. Chubb, "Managing Waste: Critical to Competitiveness," *Wasteline* (Du Pont Company), Spring 1986.

83. See OTA, *Serious Reduction of Hazardous Waste* (Washington D.C.: U.S. Government Printing Office, 1986); Sarokin et al., *Cutting Chemical Wastes*; Donald Huisingh et al., *Proven Profits from Pollution Prevention: Case Studies in Resource Conservation and Waste Reduction* (Washington D.C.: Institute for Local Self-Reliance, 1986).

84. Kirsten U. Oldenburg and Joel S. Hirschhorn, "Waste Reduction: A New Strategy to Avoid Pollution," *Environment*, March 1987; Geiser et al., *Foreign Practices in Hazardous Waste Minimization*; "How Sites Are Tackling Hazardous Waste," *Wasteline* (Du Pont Company), Spring 1986.

85. Scott Teets, General Manager, Worksafe Products Division, Brulin & Company, Indianapolis, Ind., private communication, May 28, 1987; Brulin & Company, information packet on Worksafe products; "Motion Submitted by the Greens' Parliamentary Group Relating to Policy on Chemicals," unpublished paper provided by author, Green Party science advisor Hanswerner Mackwitz, Bonn, West Germany, undated.

86. Donald Huisingh and John Aberth, "Hazardous Wastes: Some Simple Solutions," *Management Review*, June 1986.

87. Oldenburg and Hirschhorn, "Waste Reduction: A New Strategy"; Sarokin et al., *Cutting Chemical Wastes*; EPA, "Waste Minimization Findings and Activities," *Fact Sheet*, Washington D.C., October 1986.

88. Christiansen and Hansen, "Resource Recovery of Hazardous Wastes"; Christiansen, private communication; 1981 recycling estimates from EPA, "Waste Minimization Findings and Activities"; CBO, *Hazardous Waste Management*.

89. Geiser et al., *Foreign Practices in Hazardous Waste Minimization*.

90. Ibid.; Kim Christiansen and Tom Hansen, Technological Institute of Copenhagen, Taastrup, Denmark, private communication, February 2, 1987; Walker Banning et al., "North American Waste Exchanges: A History of Change and Evolution," in Center for Environmental Studies, Arizona State University, *Proceedings of the Third National Conference on Waste Exchange* (Tempe, Ariz.: 1986); Northeast Industrial Waste Exchange, "Users Guide for Computer-Based Waste Materials Listings Service," Syracuse, N.Y., undated.

91. Virginia Cooperative Extension Service, *National Evaluation of Extension's IPM Programs*; budget figures from C. David McNeal, Jr., IPM Program Leader, Extension Service-USDA, Washington, D.C., private communication, May 28, 1987.

92. Hansen, "Alternatives to Pesticides in Developing Countries"; G.W. Bird, "Alternative Futures of Agricultural Pest Management," paper presented at the IAA Symposium; federal research funding from estimate by

Howard Waterworth, USDA Agricultural Research Service, Washington, D.C., private communication, May 28, 1987, and from Robert C. Riley, USDA Cooperative State Research Service, Washington, D.C., private communication, May 28, 1987.

93. Pesticide sales figure from EPA, "Pesticide Industry; 1985 Market Estimates."

94. Banpot Napompeth, "Biological Control and Integrated Pest Control in the Tropics—An Overview," paper presented at the symposium Towards a Second Green Revolution: From Chemicals to New Biological Technologies in Agriculture in the Tropics, Rome, September 1986; David Greathead, Director, Commonwealth Institute of Biological Control, Silwood Park, United Kingdom, private communication, January 20, 1987.

95. Robert Repetto, *Paying the Price: Pesticide Subsidies in Developing Countries* (Washington, D.C.: World Resources Institute, 1985); Dover, *A Better Mousetrap*.

96. Marc Pallemaerts, "Export Notification," *European Environment Review* (Graham & Trotman, London), February 1987; Netherlands information from David Wirth, Natural Resources Defense Council, Washington, D.C., private communication, August 14, 1987; Jan Henselmans, "The Work of International Organizations on Existing Chemicals," in Institute for European Environmental Policy, *Protecting the Public Against Hazards from Chemicals*, Proceedings of the Second International Conference on Chemicals in the Global Environment, Brussels, January 29-31, 1986.

97. The President of the Republic of Indonesia, Presidential Instruction No. 3, "Improvement of Control of Brown Planthopper (Wereng Coklat), An Insect Pest of Rice," Jakarta, Indonesia, November 5, 1986; "Against the Grain in Indonesia," *Asiaweek*, March 22, 1987; training target from Hansen, private communication.

98. Vibeke Bernson, National Chemicals Inspectorate, Solna, Sweden, private communication, February 5, 1987; Jesper Kjølholt, Centre for Terrestrial Ecology, National Agency of Environment Protection, Copenhagen, Denmark, private written communication, August 6, 1987.

99. Governor Madeline Kunin, "Pesticide Policy Statement," May 14, 1986 (made available by Vermont Public Interest Research Group, Montpelier, Vt.); Dave Jensen, Nebraska Department of Environmental Control, Lincoln, Nebr., private communication, May 27, 1987; Bernard Hoyer, Iowa Department of Natural Resources, Geological Survey Bureau, Iowa City, Iowa, private communication, June 10, 1987.

100. Katherine Reichelderfer, Associate Director, Resources and Technology Division, Economic Research Service, USDA, Washington, D.C., private written communication, July 8, 1987.

101. Subtitle D, "Conservation Reserve," of the U.S. Food Security Act, *Congressional Record—House*, December 17, 1985; Michael R. Dicks and Katherine Reichelderfer, "Choices for Implementing the Conservation Reserve," Agriculture Information Bulletin No. 507, USDA Economic Research Service, Washington, D.C., March 1987; Michael Dicks, "More Benefits with Fewer Acres Please!" *Journal of Soil and Water Conservation*, May/June 1987; bill to amend the Food Security Act of 1985, introduced into the U.S. Senate, July 1987.

102. Florence Petillot, "The Policies and Methods Established for Promoting the Development of Clean Technologies in French Industry," *Industry and Environment*, October/November/December 1986.

103. Piasecki and Davis, *Lessons from Europe*.

104. Klaus Müller, Ministry of the Environment, National Agency of Environmental Protection, Copenhagen, Denmark, private communication, February 2, 1987; Dr. Stolz, Ministry of the Interior, Bonn, West Germany, private communication, January 27, 1987; Eberhard Meller, Federation of German Industries, Cologne, West Germany, private communication, January 28, 1987.

105. EPA, *Report to Congress: Minimization of Hazardous Waste*; OTA, *From Pollution to Prevention: A Progress Report on Waste Reduction* (Washington, D.C.: U.S. Government Printing Office, 1987).

106. Description of legislation from U.S. Representative Howard Wolpe, "The Hazardous Waste Reduction Act," *Congressional Record*, Washington, D.C., June 26, 1987; OTA, *From Pollution to Prevention*.

107. OTA, *Serious Reduction of Hazardous Waste*; Roger N. Schecter and Gary E. Hunt, "North Carolina's Pollution Prevention Pays Program and an Overview of Other States' Waste Reduction Efforts," N.C. Department of Natural Resources and Community Development, Raleigh, N.C., 1986; Minnesota Technical Assistance Program, information packet, University of Minnesota, Minneapolis, Minn., undated; Robert Ludwig, Alternative Technology Section, Department of Health Services, Sacramento, Calif., private communication, June 3, 1987.

108. Garland, "Report on Consultancy to the Republic of Korea"; U.N. Environment Programme, *International Symposium on Clean Technologies: Synopsis of the Country Reports* (Paris: 1986); Dhira Phantumvanit and Suthawan Sathirathai, "Promoting Clean Technologies in Developing Countries," *Industry and Environment*, October/November/December 1986.

109. Mogens Palmark, ChemControl a/s, Copenhagen, Denmark, private communication, February 3, 1987; ChemControl, "Dealing with Hazardous Waste Problems: Statement of Qualifications," Copenhagen, Denmark, undated; Whitman Bassow, "Major Corporations to Train Indonesian Officials in Industrial Environmental Management," *Environmental Conservation*, Summer 1986.

110. See J. Clarence Davies, "Coping with Toxic Substances," *Issues in Science and Technology*, Winter 1985; Carl Pope, "An Immodest Proposal," *Sierra*, September/October 1985.

111. "Restrictions on Toxic Discharges into Drinking Water; Requirement of Notice of Persons' Exposure to Toxics," Proposition 65 description in literature distributed to California voters, 1986.

SANDRA POSTEL is a Senior Researcher with Worldwatch Institute. She is author of Worldwatch papers on water conservation, air pollution effects on forests, and changes in the earth's chemistry, and coauthor of *State of the World 1987*. She studied geology and political science at Wittenberg University and resource economics and policy at Duke University.

THE WORLDWATCH PAPER SERIES

No. of Copies*

______ 1. **The Other Energy Crisis: Firewood** by Erik Eckholm.
______ 3. **Women in Politics: A Global Review** by Kathleen Newland.
______ 7. **The Unfinished Assignment: Equal Education for Women** by Patricia L. McGrath.
______ 10. **Health: The Family Planning Factor** by Erik Eckholm and Kathleen Newland.
______ 13. **Spreading Deserts—The Hand of Man** by Erik Eckholm and Lester R. Brown.
______ 16. **Women and Population Growth: Choice Beyond Childbearing** by Kathleen Newland.
______ 18. **Cutting Tobacco's Toll** by Erik Eckholm.
______ 21. **Soft Technologies, Hard Choices** by Colin Norman.
______ 25. **Worker Participation—Productivity and the Quality of Work Life** by Bruce Stokes.
______ 27. **Pollution: The Neglected Dimensions** by Denis Hayes.
______ 28. **Global Employment and Economic Justice: The Policy Challenge** by Kathleen Newland.
______ 29. **Resource Trends and Population Policy: A Time for Reassessment** by Lester R. Brown.
______ 30. **The Dispossessed of the Earth: Land Reform and Sustainable Development** by Erik Eckholm.
______ 31. **Knowledge and Power: The Global Research and Development Budget** by Colin Norman.
______ 33. **International Migration: The Search for Work** by Kathleen Newland.
______ 34. **Inflation: The Rising Cost of Living on a Small Planet** by Robert Fuller.
______ 35. **Food or Fuel: New Competition for the World's Cropland** by Lester R. Brown.
______ 36. **The Future of Synthetic Materials: The Petroleum Connection** by Christopher Flavin.
______ 37. **Women, Men, and The Division of Labor** by Kathleen Newland.
______ 38. **City Limits: Emerging Constraints on Urban Growth** by Kathleen Newland.
______ 39. **Microelectronics at Work: Productivity and Jobs in the World Economy** by Colin Norman.
______ 40. **Energy and Architecture: The Solar and Conservation Potential** by Christopher Flavin.
______ 41. **Men and Family Planning** by Bruce Stokes.
______ 42. **Wood: An Ancient Fuel with a New Future** by Nigel Smith.
______ 43. **Refugees: The New International Politics of Displacement** by Kathleen Newland.
______ 44. **Rivers of Energy: The Hydropower Potential** by Daniel Deudney.
______ 45. **Wind Power: A Turning Point** by Christopher Flavin.
______ 46. **Global Housing Prospects: The Resource Constraints** by Bruce Stokes.
______ 47. **Infant Mortality and the Health of Societies** by Kathleen Newland.
______ 48. **Six Steps to a Sustainable Society** by Lester R. Brown and Pamela Shaw.
______ 49. **Productivity: The New Economic Context** by Kathleen Newland.
______ 50. **Space: The High Frontier in Perspective** by Daniel Deudney.
______ 51. **U.S. and Soviet Agriculture: The Shifting Balance of Power** by Lester R. Brown.

*Worldwatch Papers 2, 4, 5, 6, 8, 9, 11, 12, 14, 15, 17, 19, 20, 22, 23, 24, 26, and 32 are out of print.

_____ 52. **Electricity from Sunlight: The Future of Photovoltaics** by Christopher Flavin.
_____ 53. **Population Policies for a New Economic Era** by Lester R. Brown.
_____ 54. **Promoting Population Stabilization** by Judith Jacobsen.
_____ 55. **Whole Earth Security: A Geopolitics of Peace** by Daniel Deudney.
_____ 56. **Materials Recycling: The Virtue of Necessity** by William U. Chandler.
_____ 57. **Nuclear Power: The Market Test** by Christopher Flavin.
_____ 58. **Air Pollution, Acid Rain, and the Future of Forests** by Sandra Postel.
_____ 59. **Improving World Health: A Least Cost Strategy** by William U. Chandler.
_____ 60. **Soil Erosion: Quiet Crisis in the World Economy** by Lester Brown and Edward Wolf.
_____ 61. **Electricity's Future: The Shift to Efficiency and Small-Scale Power** by Christopher Flavin.
_____ 62. **Water: Rethinking Management in an Age of Scarcity** by Sandra Postel.
_____ 63. **Energy Productivity: Key to Environmental Protection and Economic Progress** by William U. Chandler.
_____ 64. **Investing in Children** by William U. Chandler.
_____ 65. **Reversing Africa's Decline** by Lester Brown and Edward Wolf.
_____ 66. **World Oil: Coping With the Dangers of Success** by Christopher Flavin.
_____ 67. **Conserving Water: The Untapped Alternative** by Sandra Postel.
_____ 68. **Banishing Tobacco** by William U. Chandler.
_____ 69. **Decommissioning: Nuclear Power's Missing Link** by Cynthia Pollock.
_____ 70. **Electricity For A Developing World** by Christopher Flavin.
_____ 71. **Altering the Earth's Chemistry: Assessing the Risks** by Sandra Postel.
_____ 72. **The Changing Role of the Market in National Economies** by William U. Chandler.
_____ 73. **Beyond the Green Revolution: New Approaches for Third World Agriculture** by Edward C. Wolf.
_____ 74. **Our Demographically Divided World** by Lester R. Brown and Jodi L. Jacobson.
_____ 75. **Reassessing Nuclear Power: The Fallout From Chernobyl** by Christopher Flavin.
_____ 76. **Mining Urban Wastes: The Potential for Recycling** by Cynthia Pollock.
_____ 77. **The Future of Urbanization: Facing the Ecological and Economic Constraints** by Lester R. Brown and Jodi L. Jacobson.
_____ 78. **On the Brink of Extinction: Conserving Biological Diversity** by Edward C. Wolf.
_____ 79. **Defusing the Toxics Threat: Controlling Pesticides and Industrial Waste** by Sandra Postel.
_____ **Total Copies**

Bulk Copies (any combination of titles) **Single Copy** $4.00
2-5: $3.00 each 6-20: $2.00 each 21 or more: $1.00 each

Calendar Year Subscription (1987 subscription begins with Paper 75) U.S. $25.00 ___

Make check payable to Worldwatch Institute
1776 Massachusetts Avenue NW, Washington, D.C. 20036 USA

Enclosed is my check for U.S. $ ___

____
name
____
address
____
city **state** **zip/country**